Books C...

Beginner's Bee Book

By Ted Hooper

Text © Ted Hooper, 2014.
First published in the United Kingdom, 2014,
by Stenlake Publishing Ltd.
54-58 Mill Square,
Catrine,
KA5 6RD

Telephone: 01290 551122
www.stenlake.co.uk

ISBN 9781840336214

Printed in China

**The publishers regret that they cannot supply
copies of any pictures featured in this book.**

Contents

Introduction

Beekeeping is interesting and absorbing. It captures the imagination of those who take part in a way which is unique and which leads them on to ever greater co-operation with the honey bee. There are as many reasons for starting beekeeping as there are aspects to the satisfaction that people get from the craft. The keen fruit grower, amateur or professional, may start beekeeping to provide the pollination of the blossom and the resulting setting of fruit. A keen conservationist may realise how the honey bee provides much of the pollination of wild flowers, particularly as the populations of wild bees have diminished under the pressure of the last half century's agricultural practices. Beekeeping is one of the few pastimes which can recoup the money spent on it. Indeed if the job is done properly a profit can be made. Many people eat honey and revel in its purity as one of the foods unaltered by man from its original composition. It is, however, considered by some to be expensive and they often set up in beekeeping to provide their own from their own bees. Some people get interested in the honey bee as an insect in its instinctive ability to travel around the countryside, find its way home, and then be able to tell the others where it has been. The honey bee is so far removed from ourselves in most ways but at the same time organises itself into large groupings of individuals, colonies, the organisation of which is, to us, intriguing because it does not appear to be based upon any form of leadership. Some therefore keep bees to observe, study and research while others keep them because they like them at the bottom of the garden to go down and look at, to listen to their hum and enjoy a peaceful interlude.

Once having started keeping bees then the peripheral interests become more important and more varied. You cannot be a beekeeper for more than a short while before your interest in flowers becomes more pointed and you begin to look at them as sources of nectar and pollen for your bees. Agricultural practice becomes important; not only does this control most of the large acreage of bee forage but also the need to control pests can, if done unthinkingly, destroy the bees from your hive and most of the wild insects, beneficial or neutral. In this way one is led into thinking about the wild bees and other insects so that some beekeepers become quite well informed amateur entomologists.

For the more academic there is the study of the history of beekeeping, the use of the honey bee in ornamentation, symbolism, heraldry and art, as well as the vast literature which has built up over the centuries in every language.

In short the beekeeper should never feel he has got to the end of a subject which is surprising in its depth and lateral spread but which is so simple in its practice. This book will hopefully teach aspiring beekeepers how to start with as few setbacks as possible, and solve many of the problems which all beginners face and get them going in a wonderful pastime.

1. The Honey Bee and its Colony

The colony

A colony of honey bees will consist of a queen and some 10,000 to 60,000 of her daughters, who will be the workforce of the colony and are therefore called worker bees. In the summer there will also be about 500 to 1,000 male bees, drones, the queen's sons. The queen is mother to all the bees in the colony. This large assemblage of insect will be living on the surface of the comb. The comb is made of beeswax, which the workers produce from glands in their abdomen. The beeswax is fashioned into six-sided, hexagonal, cells on each side of a central midrib. There will be a number of these vertical combs hanging side by side, and parallel to each other, the whole constituting the bees' nest. The cells are used as containers to store honey and pollen in and to act as cradles for the baby bees; baby bees are called brood by the beekeeper. Within the honey bees' nest the distribution of honey, pollen and brood will always be the same. By removing a comb a section across the nest can be seen. Honey will be in an arch at the top of the comb, pollen will be in an arch below it and the eggs and brood will be below, and central, filling an almost circular area. Pollen is sometimes scattered amongst the brood but honey is always kept over the top of the brood, extending down the sides if there is enough of it and the brood does not extend all the way across the comb. Taking out combs one after the other will show that this distribution is three dimensional with the brood in a central ball enclosed in a bell of honey. The volume of comb in which the eggs and brood are situated is called the brood nest; honey and pollen in the comb is collectively called stores.

A worker honey bee.

7

Cells

There are three main sorts of cells fashioned by the bees. Worker cells are the smallest hexagonal cells, covering about five to the linear inch, or 25 per square inch. The baby workers are raised in these cells, hence their name. Drones' babies are raised in drone cells, hexagonal cells larger than worker cells being about four to the linear inch. Queen cells are specially built when young queens are to be produced in the colony. They are quite unlike the other cells being acorn shaped and hanging vertically from the edge or face of the comb.

Life cycle

The individual bee starts off as an egg from which hatches a pearly white legless grub, the larva, which grows and turns into a pupa. The pupa goes through a complete change of form, or metamorphosis, eventually turning into the adult form, the imago or adult bee.

Eggs

The eggs of the honey bee are small translucent sausages about 1.6mm long by 0.4mm thick. They are laid by the queen glued to the base, or midrib of the cell, only one to a cell under normal conditions. All honey bee eggs develop whether they are fertilised or not. Drones are produced from unfertilised eggs, and therefore have only one set of chromosomes, those from their mother. They only carry the characteristics of their mother and are not related to the drones with which she mated. Females hatch from fertilised eggs, and have a father and mother. During metamorphosis they can turn into either queens or workers depending upon the way they are fed during their larval period by their worker bee nurses. Where a species has individuals of the same sex, which have quite distinct forms, the behaviour patterns or physiologies of these different types are termed castes. The honey bee therefore has two female castes, queens and workers. All eggs hatch in three days after being laid.

Larva

From the eggs small larvae hatch. These are the equivalent to the caterpillar of a moth or butterfly, but in the honey bee they are pearly white, legless grubs curled up on the base, the central midrib, of the cell. The larvae are fed by the workers with a sort of bee milk, produced in glands in their heads, mixed with some honey. This mixture is termed brood food or worker jelly when fed to worker larvae and royal jelly when fed to queen larvae. Different proportions of the constituents of the food are used when the bees feed a worker or a queen larva and these differences trigger the control of hormones in the larva which cause it to develop into either a queen or a worker. This food is extremely nutritious and easily absorbed by the

larvae so that growth is very rapid. In about five days each larva fills the bottom of its cell. The workers seal them in by placing a domed wax capping over each cell. On worker brood the cappings are slightly convex but on drone brood they are more pronounced, quite large and high; this stage is called sealed or capped brood. Queen cells are capped over by sealing the open end. After capping the larva spins a light cocoon around the inside and lies longwise in the cell. Queen larvae only spin their cocoon over the lower half of the cell. After spinning its cocoon the larva turns into a pupa. This stage is immobile but internally very active as it reforms its larval body into that of the adult bee. At the end of this period of metamorphosis the bee chews its way through the capping and emerges as a fully developed individual equipped with all the instincts required to take part in the work of the colony. The three types of individual pursue their metamorphosis at differing rates, as shown in table 1. These life cycle periods should be committed to memory as they have considerable importance in the practical management of colonies.

Table 1. Timetable of Honey Bee development – life cycle in days from the laying of the egg.

Type of bee	Queen	Worker	Drone
Egg hatches	3	3	3
Cell capped over	9	9	10
Bee emerges	15/16	21	24
Death	1–3 years	36 days	90 days

These development times can vary according to the race of honeybee and the temperature.

New comb containing eggs.

Curled larvae, some being fed by workers.

Behaviour

The adult lives of the three types of bee are totally different but all are equally important in the life of the colony. Honey bees emerge with all the instincts necessary to do all the complex jobs required by the colony for its establishment, maintenance and survival. The details of the organisation and internal controls of a bee colony are outside the scope of this book but is a subject well worth pursuing for those interested. Here we will deal only with the patterns of behaviour which are of direct use in the management of colonies.

Drone behaviour

It is not known if the drone bee does any work in the hive. As far as we know, the drone's only role in the colony is to fly out to mate with new young queens. Drones are mature and ready to mate about 14 days after emerging from the sealed cell. They fly out on mating flights only when it is reasonably warm and calm. The rest of the time they stay indoors being fed by the workers or sleeping quietly, usually several together. When they have been out on a flight they appear to be welcomed into any colony as well as their own.

Mating takes place high in the air. In some areas drones assemble in specific mating areas and in others they are randomly spread around the sky. Drones are attracted to the queen by scent and die at the time of copulation. Drones have been known to fly for nine or ten miles from their hives to mating areas and are therefore important as one of the main means of gene dispersal in the species. The workers kill the drones, mainly by starvation, at times of extreme lack of honey during the summer and, almost always, at the end of the season before winter in those areas where cold winters occur.

Queen behaviour

The queen is mature about four days after emergence and usually starts laying in about 10 to 20 days. Sometime during this period, when the weather is calm and warm, the queen flies out to mate with the drones. She mates with from five to fifteen drones on one to three mating flights. By three weeks after emergence she is usually incapable of mating successfully. The sperms received from the drone during mating are stored in a special organ, (spermatheca), in her body and used during the rest of her life to fertilise the eggs as each is laid. She never mates again. It will be seen from this that although the queen is mother of all the members of the hive the worker honey bees are not all full sisters, many of them are half-sisters to each other, having different drones as fathers. Thus the observed characteristics of a colony may vary slightly from time to time. Once the queen starts to lay this becomes her full-time job and a good queen will produce around 2,000 to 2,500 eggs per day at the peak period of laying. The queen has considerable influence on the behaviour of the colony through the production of chemical substances (pheromones), which affect the physiology and behaviour of other individuals in the colony.

Worker behaviour

The worker honey bees do all the work required to maintain the colony. Therefore, unlike the queen and drone which have a restricted set of instincts, they are equipped with a wide spectrum of instincts and drive to do the various jobs necessary in the colony, and with the innate ability to learn from their contact with the environment where this is crucial to their well-being.

House bees

The lives of worker bees are roughly divided into two main periods. During the first 15 to 16 days after they emerge they are house bees or nurse bees and work in the colony cleaning, building comb, feeding larvae, capping larvae, guarding the colony, accepting nectar from bees working in the field and processing it into honey for storage. During this period they may take flights on most days if the weather is nice, but they do not collect anything in the field or bring anything home. During these flights they learn the position of the hive in relation to local landmarks – bushes, trees, buildings and general landscape and the rate the sun moves across the sky, which is of great importance to their method of navigation. Towards the end of their period as house bees they come into contact with the bees bringing back nectar and pollen from the field. These bees do a small dance which indicates to other bees where they have been to collect the substance they have brought back. The dance indicates the direction and distance to the source of forage and the dancers provide a sample which acquaints the bees following the dance with the taste and smell of the substance. The older house bees get involved in following these dancers and eventually become attached to one dance, hence to one type of flower in a particular area of their flight range. They then fly out to collect and bring back the same substance. In this way they become foragers and spend the rest of their life working in the field bringing in the necessary provisions for the colony. They do no further work indoors under normal colony conditions, although they can revert to house duties if the good of the colony requires it.

Forager behaviour

Honey bee foragers collect, and bring into the colony, four substances: nectar, water, pollen and propolis. The first two are swallowed into the honey crop, which acts as a storage container where it is held until it is regurgitated to house bees in the hive. The other two are conveyed into the hive on the hind legs packed in the pollen baskets or corbicula. There are hairs on the back legs which support the loads being brought in, there being no true basket or pouch.

Nectar

Nectar is a sugar solution collected from flowers. The concentration of the sugar is very variable but normally about 20%–60%. The house bees turn this into honey for storage by evaporating some of the water, reducing it to about 18%–20%, and breaking up any complex sugars into simple component sugars, mainly glucose and fructose. This process, which raises the concentration of the sugar to some 80%, prevents the honey fermenting during storage.

Honey

Honey is the energy food of the honey bee. It is used to fuel its bodily functions, including flying, and the production of heat to warm the colony. When comb is to be built honey is eaten and its sugars converted into beeswax in the wax glands of the house bees.

Water

Like all living things the honey bee needs water for its bodily functions but only a few bees collect water to bring back to the hive, where it is used to dilute stored honey for use by the bees. Most bees obtain their water in the nectar or diluted honey which is circulating amongst the bees at all times. Bees constantly feed each other, so much so that the house bees at least have the same substances in their stomachs. This helps in the distribution of some pheromones and in the production of a unique colony odour. Water is also used to cool the hive down. The bees fan to evaporate the water, see page 18.

Pollen

The honey bee enters flowers and collects pollen from the stamens – the male part of a flower. The pollen is dusted all over its body. Using all her legs she will brush the pollen from her body adding a little nectar to stick the grains together, and packing the result into the pollen baskets, carry it back to the hive, where it is pushed off into an empty cell, or one that already has some pollen in it. House bees then push it down flat with their heads and it is eaten as required. Pollen is the main food of the bee. It provides the bees requirements for protein, vitamins and minerals. Bees need to eat pollen before they can produce bee milk or brood food.

Propolis

Propolis is 'bee glue'. It is a sticky, resinous, substance collected from plant buds and packed into the pollen baskets. House bees chew it from the foragers' legs and use it to fill holes in the hive, glue down anything loose and to erect draught curtains in the entrance to the colony. This use of propolis to glue everything together in the hive is the main reason for wearing gloves. Gloves prevent the fingers becoming sticky. One can always take of the gloves to do delicate jobs such as handling the queen, without the fear of getting her covered in the sticky substance.

Queen replacement

Queens live on average for about two to three years so the colony must have a means of replacing her. At the same time it would be wasteful for the colony to be producing queens all the time in case they were needed. There is therefore a mechanism which will trigger the bees to produce new young queens when necessary. Basically the trigger is a pheromone, known as 'queen substance', which is produced by the queen and licked from her body by the workers who are attending to her needs. Queen substance is passed from bee to bee during food sharing and while the amount in circulation is above a certain threshold the worker bees are inhibited from making queen cells. Should the quantity of pheromone fall below the threshold, either because the queen is ageing and is not producing a sufficient amount or the colony is so congested that the sharing of the pheromone breaks down, then queen cells are built.

In these two cases the bees construct queen cell cups and the queen lays in them, after which the bees nurse them to adulthood. Once new young queens are produced the colony can supersede, replace the queen with a new one, or swarm, replacing the queen of the colony and at the same time establish other colonies with some of the workers and young queens, or they may give the process up and kill the young queens. From the practical point of view one cannot rely on the latter happening as it is impossible to forecast when the bees will follow this course. The pheromone could suddenly vanish if the queen dies or is killed by the beekeeper, accidentally or on purpose. Queen cells produced when the queen is lost or killed unexpectedly are termed emergency cells and the resulting queens emergency queens.

Emergency queen cells

When the queen has died, or been killed, there is no one to lay eggs in the queen cups so the bees have to use another method of queen cell production. In this case the bees select several very young larvae and modify their cells to expand the cell mouth and from it turn a queen cell downwards on the face of the comb. Royal jelly is added until the larvae are floated up into the right position in the vertical queen cell. They are then nursed on to adulthood when one is elected and takes over the colony. Unfortunately the bees will often make emergency cells on larvae which are too old to make good queens, so from the practical viewpoint emergency queens should not be used to head the beekeeper's colonies, unless produced as part of simple queen rearing. See 'Queen Replacement – production of' page 116.

Supersedure

When queen cells are seen there is no certain way of knowing whether they are being prepared for supersedure or swarming. The supersedure queen cells are likely to be few in number, four or five, and half way up and in from the sides of the comb. However, this is not a reliable diagnosis. In fact it is probable that colonies in the middle of supersedure can swarm. These conditions are seen towards the end of the season, or early in the year, supersedure may sometimes be induced by reducing the queen cells to one while leaving the old queen to carry on. Most supersedure is accomplished by the bees without the beekeeper being aware of it. If the beekeeper's queens are not clipped or marked it will probably never be realised that supersedure has occurred. In a number of cases of supersedure the old queen is not killed by the bees and one finds the mother and daughter both laying in the colony, often on the same comb face. Eventually the old queen disappears and the supersedure queen can be left to head the colony.

Swarming

This is the process whereby the colony re-queens itself and at the same time splits up to form several more colonies. It is colony reproduction. Swarming has become necessary to the honey bee because the queen, unlike the queen bumble bee or social wasp, is quite incapable of setting up a new colony on her own as she cannot build comb or feed her young.

A number of queen cells are produced, often a dozen or more, and on average the old queen will leave with the first swarm soon after the first queen cell is sealed. Usually the swarm, consisting of 10,000 to 20,000 workers and the old queen plus a few drones, leaves the hive having filled up with about three days' supply of honey. They cluster close by in a tree or on a post. Scouts will be sent out to find a new home and, once the decision has been made, the swarm takes to the air once more and may fly several miles to their new home. Meanwhile back in the colony the worker brood is emerging and the adult population is building up, while the amount of brood is reducing, there being no new eggs to develop, Several of the queen cells may be ready to emerge and the workers will allow one to come out and hold the others in by physically holding down the capping on the cell and resealing it as the queen tries to cut her way out. The first young queen will leave the hive with the first after swarm or cast. Two to three days later one or two more young queens will be allowed to emerge and will go off with another cast, this process being repeated until the colony decides to finish swarming, at which time a young queen will be accepted as queen to head the colony and any others will be killed. The workers are quite aggressive towards the new queen until she has mated and starts laying. It can be seen from this description that left to themselves colonies can reduce their population certainly below that required to produce the beekeeper a surplus of honey and often below the level at which they will survive.

Collecting a swarm in a skep.

Winter survival

Honey bees evolved in the tropics and have not reduced their temperature requirements as they have moved north into colder regions. Brood must still be kept at about 95°F/35°C, and adult bees have difficulty working below 60°F/15.5°C. The bees can produce heat in their bodies by metabolising the sugar from honey in the large wing muscles of the thorax. This happens automatically during flight but when the bees are in the hive they can raise their temperature by shivering, flexing the wing muscles to produce the heat required. The honey bee has evolved two main methods of dealing with the winter in the colder areas of the world. First it reduces brood rearing or stops entirely, and secondly conserves the heat produced by the individual bees by clustering. As the ambient temperature reduces the cluster packs together tighter, thus reducing the loss of heat by convection in the air between the bees and makes its collective surface area smaller, thus reducing radiation from the cluster as a whole. The bees do not heat the hive, only the area where the cluster is. In this way bees can deal with all but the very cold arctic conditions. In warmer areas brood rearing is more or less continuous and colonies never go into deep cluster. If the temperature rises above that required by the brood the bees will first ventilate the cluster by flapping their wings like small fans called fanning. If this is insufficient they fetch in water and evaporate this to cool the combs, brood and cluster.

2. Starting Beekeeping

To start beekeeping a new beekeeper will need to get a nucleus colony or swarm of honey bees and an apiary site ready to take them. The new beekeeper will also need personal equipment, a hive and ancillary equipment.

Personal equipment, clothes

When manipulating honey bees it is necessary to wear clothes which are smooth so that bees alighting on them do not get their feet tangled as doing so irritates them and may cause them to sting. It is best to wear a white boiler suit made of cotton or nylon. These can be obtained with built-in veils which are very bee-proof and easy to use.

Veils

It is absolutely necessary to wear a veil when handling bees to prevent stings on the face, which are painful and quite unnecessary. Veils come in many sorts. They can be built into overalls, as mentioned above, or separate. The main thing is that they should be made bee-proof and constructed in a way which prevents the veil from touching the face when it is windy, or the back of the neck when bent over working.

Examining a broodnest.
Beekeeper kneeling to prevent backache.

Gloves

I think that it is best for a beginner to wear gloves when first starting. The beginner will put a pair of gloved hands down onto the bees with greater confidence than without. Confidence is the essence of bee handling. Lack of confidence causes most of the stings suffered by beginners. Gloves also prevent the hands getting sticky with propolis. Later on, when confidence has been gained, the gloves can be taken off and one has nice clean hands to handle individual bees or queens. It is well worth getting proper purpose-made bee gloves which, although expensive, are the most efficient in my opinion. For hygiene reasons disposable gloves are now advocated. Certainly when working other person's bees, disposable gloves should be worn. The disadvantage is that most person's hands sweat in disposable rubber, vinyl or nitrile gloves.

Manipulating equipment

Hive Tools

Hive tools are used to separate the parts of the beehive and to loosen frames for inspection. There are three main sorts, of these I prefer the one with wide flat blade which does the least damage to the hive when separating boxes and a wide flat hooked end to give good leverage when used to separate frames from each other. Hive tools should be painted a bright colour so that they can be seen easily if dropped into long grass.

Smoker

When bees are to be manipulated they are first given a few puffs of smoke. This causes them to fill themselves with honey making them more docile and easy to handle. A smoker is therefore an essential piece of equipment. The bent-nosed type as illustrated is best, the tiny straight smokers are fairly useless and more difficult to keep alight. The smoker should have a stout hook (a single coat hook is suitable) screwed to the back of the bellows so that it can be hooked onto the side of the hive while manipulating a colony. This keeps it close to hand and you do not have to bend down to pick it up every time it is required. If you do put the smoker on the ground stand it upright. If you lay it on its side it will quickly go out.

Bent nose smokers with flat blade hive tools attached for safe storage.

Beekeeping equipment

The new beekeeper will need a hive comprised of a floor (open mesh or varroa floor is now recommended) and entrance block, a brood chamber, a queen excluder, three supers, a crown board, a roof and the frames to fit. Foundation will be needed, but for a start enough for a brood chamber and one super will suffice. Necessary ancillary equipment will be a feeder, a clearer board, a hive stand and if possible a couple of dummy frames. DIY beekeepers can make most of it themselves if they wish, but it must be made accurately to standard size and design or they will be in for trouble with the bees in the future after they have glued them up with propolis. The vibration caused by un-sticking the parts of a badly made hive will annoy the bees and probably gain the beekeeper a few stings. The newcomer to the craft will not realise the problems of various hive designs, and most experienced beekeepers will usually be happy with what they are already using.

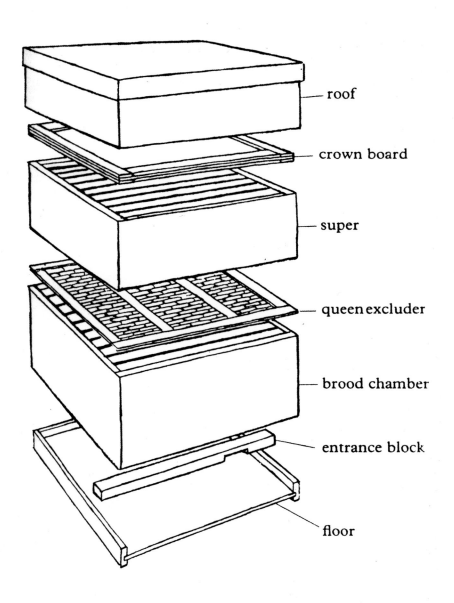

roof

crown board

super

queen excluder

brood chamber

entrance block

floor

Illustration of a basic hive.

Hives

There are innumerable types of hive and each has its own protagonists. Honey bees can be kept successfully in any of them, the bee is a very adaptable little insect, but some hives are more easily used by the beekeeper and are therefore the best buy. The design of the hive is much more important to the beekeeper than to the bees. Hives are basically a number of open ended boxes piled one above another to the size required, with at the bottom a floor, the box, and on top an inner cover or crown board, topped by an extra large cover or roof to keep the rain out (see illustration on page 21). Inside the boxes are a number of wooden frames into which the bees are persuaded to build their combs. The names of the parts of a hive are given in the illustration. The large box is called a brood chamber because it usually contains the queen and her brood. The shallow box is called a super because it is put on the top of the brood chamber. It is used to contain the combs of honey; further boxes are put on as required by the colony. The queen is usually prevented from leaving the brood chamber and entering the supers by a queen excluder, a grid through which the workers can pass but not the queen. Sometimes a second brood chamber is placed below the excluder and the queen is allowed to lay in both of them. This is spoken of as a double brood chamber colony. In other cases the size of the brood chamber is extended by placing a super above the brood chamber and below the excluder. In beekeeping jargon this arrangement is called brood and a half.

The beeway

All modern hive design is based upon the 'beeway' as first used by Langstroth in about 1852. A beeway, or bee space, is left between any two surfaces in a hive and is not joined or blocked, but used by the bees as a channel to walk through. The beeway is a space of between ¼ to ⅜ inch (6 to 10mm). If the space is reduced below 6mm the bees will fill it up with propolis, if it exceeds 10mm then they will build in a thin strip of comb.

Top or bottom beeway hives

There must always be a beeway between the bottoms of the frames in one box and the top bars of the frames in the box below. This bee space is built into the design of the box and can be formed by the sides of the bottom box projecting a beeway above the tops of the frames, called top bee space, or the sides of the box can be flush with the tops of the frames and the beeway be built into the top box under the frames, bottom bee space, see illustration. Top beeway is by far the best when manipulating, making it much easier to avoid crushing bees when putting boxes one above the other, particularly when they are heavy with honey. Unfortunately, traditionally, most British hives are constructed with bottom beeway but all can be converted to top beeway when they are being assembled.

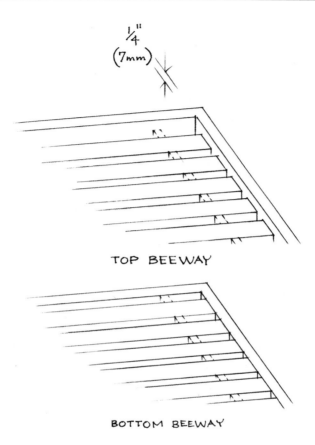

$\frac{1}{4}$"
(7mm)

TOP BEEWAY

BOTTOM BEEWAY

Buying a hive

New hives are usually bought in the flat and put together by the beekeeper. Second-hand hives may be found but be careful that they have not been in contact with either of the foul brood diseases. In any case it is best to run a blow lamp, or gas torch, over second-hand wooden equipment to kill any possible infections. Traditionally hives in Britain have been made of western red cedar but recently the cost of this material has gone up so much that now many are being made of pine. Weight of the hive is the only disadvantage and this is not a problem unless one is handling large numbers. Cedar hives will last longest but, looked after properly, pine will last a lifetime. Beekeepers can make their own hives. It is not a difficult piece of joinery but the hives must be made accurately to the standard designs to fit the frames you will use in them. The latter are usually bought in from the equipment firms, as only the most enthusiastic handyman will consider making his own.

Inaccurately made hives will cause untold trouble when making inspections and are conducive to stirring up the bees to stinging levels.

Type of hive

What type of hive should one buy? Certainly a single wall hive not one, like the WBC hive, with two sets of boxes one inside the other. It may be prettier but it is more expensive and less efficient in use. My own preference is for the Modified Commercial hive but I would accept the Langstroth hive in those areas where this is the usual hive in use. The commonest hive in Britain is the British National hive. It is a small hive which is usually used with a super as an extension to the brood chamber – brood and a half – this arrangement has more disadvantages than advantages in use, and is mainly perpetuated by tradition.

Frames

Each hive is designed to hold frames of a definite size and it is essential to obtain the correct ones when buying them. The deep frames for use in the brood chambers are usually called brood or deep frames while those for the super are spoken of as super frames or shallows. Brood frames, irrespective of the hive being used, should have top bars 27mm (1 $\frac{1}{16}$ inch) wide and Hoffman self spacing side bars. Super frames can be spaced with castellated runners. If you are using a tangential extractor or if you have a radial extractor the more efficient Manley type frame can be used see 'Extractors' page 94 – for an explanation of difference between types of extractor.

Langstroth hive

This is the hive used over most of the world. It is a single walled top beeway hive and very easy to construct oneself. It is made to take different numbers of frames in different parts of the world; it is best to enquire what the usual size is in your area. In Britain it is made to take ten Langstroth frames, but the hive is used by very few beekeepers, which is why I prefer to stick with the Commercial hive which is about the same volume – the same size from the bees point of view.

Modified commercial hive

This is a single walled hive and should be purchased as a top beeway hive. They are generally made with bottom beeway unless top beeway is asked for. It is a very simple hive for home construction each box being made out of only four pieces of wood, as is the Langstroth hive. It is made to take eleven commercial, commonly called '16 by 10', frames. In actual fact twelve frames can be put into the hive when it is new but are too difficult to get out once the bees have added propolis to glue them down. The plan size of the hive is the same as the national hive so floors and queen excluders are in common with the national hive, thus giving some economy.

British Modified National Hive

This is a single walled hive with bottom beeway. It takes eleven British Standard frames, DN/5 being the best for the brood chamber, SM/2 for the supers with castellated runners and SN/7 Manley frames. These frames have long lugs and therefore to accommodate the lugs the construction of the hive has to be more complicated than either of the above two. The brood chamber of this hive is 25% smaller than the other two mentioned above.

Smith hive

From the bees view point this hive is the same size as the British National but it contains eleven British Standard short lugged frames, types as for National but numbered DS/5, SS/2 or SS/7, and therefore each box is made of four pieces of timber, American style, as are the first two mentioned above. It is also a top beeway hive. It is very popular in Scotland where it was designed by Willie Smith, hence the name.

Wax foundation

To persuade the bees to make their comb in the wooden frames a sheet of beeswax, on which is embossed the outline of the cells, is inserted into each frame. The bees then draw out the foundation to form the cell walls, adding wax of their own, to complete the comb. Wax foundation tends to sag and buckle under the weight and heat of the bees and has to be strengthened using wires embedded in the wax.

Beginners should buy wired foundation, which is inserted in the frames as shown in illustrations. The two bottom bars should never be nailed together as the wax must be free to expand downwards when heated by the bees, sliding through the space between the bars, or deformed comb will result. The wax foundation put into super frames should have the pattern of worker cells on it.

Foundation in a wooden frame.

Dummy frames

Dummy frames, usually spoken of as just 'dummies', are made exactly the same size as the frame used in the hive but are made of a solid piece of timber about ½ inch thick. They are extremely useful pieces of equipment and beekeepers are well advised to have at least one per hive and a number of spares. They hang in the hive in the same way as frames with a beeway around the sides so that they can be easily taken in and out. Division boards, which look the same as dummies, are made without the beeway. They fit tightly to the side of the hive, and are rapidly glued in by the bees with propolis so they are almost impossible to get out. Division boards are useless. Don't waste money on them.

A frame feeder which will take syrup to feed bees, but can act as a dummy frame.

Crown boards

Crown boards, or inner covers, are made to fit on top of the boxes under the roof. The usual type for use with bottom beeway hives is made from three ply with a frame which provides a beeway on each side of the board, two large slots are in the centre and act as holes through which the bees can be fed or to take Porter escapes when the board is used to take the honey off, see 'Clearer boards' page 90. For top beeway hives crown boards do not need a frame and are best made from heavy ½ inch ply. It can have a small round 1 inch feed hole in the centre if you wish but there is little sense in making them with slots to take escapes, as will be seen when dealing with taking off the crop.

Glass quilts

Glass quilts are made the same as crown boards but glass or Perspex takes the place of the three ply and they are made with a small feed hole in the centre. For top beeway hives they need to be framed on one side only or the bees will build comb, brace comb, between the top bars and the glass. The glass quilt is used in place of the crown board and has the advantage that it is possible to look in at the bees without disturbing them. They are very useful to beginners who always want to see how their new bees are getting along and open up the colonies far too often in the initial weeks.

Queen excluders

Queen excluders are metal grids and are all made with slots which are wide enough for the workers to pass but too narrow for a normal queen to get through. They come in several types. The Waldron types are made with parallel wires forming the slots. If this sort is bought you must ensure that it is constructed so that there is a beeway only on one side; some were made with a beeway on both sides and are useless as they encourage the bees to build comb through the wires. Short or long slotted zinc sheet excluders are available. The short slot zinc excluder is my favourite as it seems to wear well, but for efficient use needs to be framed, as shown in illustration on page 21. Slotted excluders made of stiffer materials than zinc are available and are excellent but also need to be framed for use.

Queen excluder.

Feeders

It will be necessary to have some means of feeding sugar syrup to the bees for winter stores, and to prevent them from starving in a very bad spring or summer.

Miller feeder

A Miller type feeder, preferably the Rowse feeder with the feeding station at one side of the box and entry for the bees into the main box when they have taken most of the syrup so that they can clean it up – see illustration, is the best for general use. This type of feeder is placed on top of the hive, under the crown board, and the bees come up over the first wooden wall and down to the syrup, fill themselves and return down to the colony to store the syrup. When they have almost cleared the syrup they can get under the second wooden wall and clean up the whole box of any syrup left.

Miller feeder. The bees will access the food through the gap as described in the text.

Round metal feeders

This type of feeder is quite popular but is not large enough to give a couple of gallons of syrup at a time and thus slows down feeding and increases the number of visits by the beekeeper to get the job done quickly especially in the autumn when feeding down for winter.

Bucket and home-made feeders

Bucket feeders are made with a fine gauze patch inserted in the lid and are filled with syrup and inverted over the feed hole of the crown board. Similar feeders can be made from a lever lid tin of which the lid has had a patch of holes made in its centre. The patch is usually about 2 to 3 inch in diameter and the holes are punched from the outside inwards with about a ¾ inch wire nail, leaving holes of less than $\frac{1}{16}$ inch diameter. This type of feeder should always be full before placing on the hive. Using a spare bucket invert the feeder over the bucket to allow syrup to drip out until a vacuum is created. It is now safe to put the bucket on the hive with the holes directly over the feed hole of the cover board. An empty super or brood chamber should surround the feeder and a solid board placed on this to make the roof fit snugly. This type of feeder is useful as bees have access to the syrup without leaving the warmth of the brood chamber. A cold night followed by a warm morning there will be a rapid temperature rise, the air in the top of the inverted tin will expand and push a greater quantity of syrup out than the bees can deal with. I have seen syrup

Clive de Bruyn pouring syrup into a Miller feeder.

running from the hive on warm mornings. This is not only wasteful but is conducive to robbing; bees try to get into each others' hives to steal the honey, which should be avoided at all costs.

Buying bees

General

Bees should be bought from an accredited supply firm or from a local beekeeper of repute. You should specify that you want nice quiet, easily handled bees. I would advise you to contact your local beekeeping association, or beekeeping adviser, and seek their advice on where to purchase. There are four ways of getting your first colony of bees these are: buying a full colony, buying a nucleus (a small colony), buying package bees or collecting a swarm.

Buying a colony

I would never advise a beginner to start with a fully developed colony, there are too many possible problems. An established colony is an awesome sight to a beginner, who does not know what the bees are doing or what to do to stop them doing it. It could put you off beekeeping for the rest of your life. Also it's like buying a second-hand car – caveat emptor – you do not know what you are going to end up with.

Buying a nucleus

A nucleus, or 'nuc', is a small colony of bees usually covering four or five frames. It will contain a young queen, about 5,000 worker bees, eggs and brood in all stages filling at least half the cells of the comb and a couple of pounds of stores – honey and pollen – to see them through to their new home. The nucleus should contain quiet bees. Being small in number they will be very easy to handle and the beginner's experience will grow as the colony expands and grows.

Package bees

In some parts of the world the usual way to start beekeeping is with a package of bees, that is about 3 lbs of bees and a queen, no comb or brood, transported in a wire mesh container and provided with a can of syrup or a lump of candy to prevent them from starving on the journey. These again should be nice quiet bees and are small enough to make handling easy.

Swarm

The only way to obtain bees free is to collect a swarm and hive it yourself. If you are happy to do this then you must realise that there are no guarantees what sort of a colony you will get; the bees may be quiet and well-behaved or quite frisky and need little encouragement to sting. They may be disease free or infected with one or more of the major diseases. It is a matter of luck, but as long as you are aware of the position and willing to risk it then carry on and my best wishes go with you. A lot of good beekeepers started this way.

It is advisable to collect your first swarm with an experienced beekeeper.

Apiary preparation

Apiaries should always be sited so that they will not cause a nuisance to anyone and preferably well-concealed. Even in built-up areas it is unusual for bees to cause problems unless their presence is known. Many so-called problems are in the mind of the complainant rather than a matter of fact. Colonies are best sited facing a hedge and about four feet away. This provides shelter and calm air at the front of the hive where the bees will be decelerating to land. The colony is best on the sunny side of the hedge and if possible the site should not be in a frost pocket at the bottom of a slope. If it is likely other people will be around then the hedge can be grown up to about seven feet in height which will push the foragers up above head height. Bees do not usually sting anyone unless the colony is being manipulated, but will sting if they become accidentally tangled in someone's hair. Inside the apiary the hives should be placed about six feet apart and each facing a slightly different way to prevent young bees drifting into the wrong colony. Hives are best placed on stands at a convenient height for the beekeeper. I prefer to have the top of the brood chamber at knuckle height when standing by it. At this height one can stand relaxed, not bending over, when taking frames out for examination and the frames can be removed without rubbing bees against each other or crushing them against the side of the box, something to which they and their friends take great exception.

An apiary laid out in a circle.

An apiary laid out in a straight line with the entrances facing in different directions.

3. How to Examine a Colony

Working with the colony

Dressing

Before touching the bees you should put on your veil and overalls and make sure that everything is fitting properly and bee proof. It is hard to believe as a beginner that bees only sting if you stir them up. You feel that every bee that lands on you is intent on doing you harm and do not realise that the poor thing has landed on you wondering what this new white apparition is which has suddenly appeared at the side of the hive and so changed the local landmarks. Good protective clothes give you confidence until you begin to understand bees and recognise their behaviour.

Lighting the smoker

The next thing to do is light the smoker. Do this before putting on your veil, the black netting is very flammable. All sorts of things are used as smoker fuel. I prefer dried grass because it burns slowly and the smoke is reasonably innocuous when you inhale a lung full of it. Usually the grass is cut with a rotary mower and once a year this is done leaving the grass box off, and the cuttings are left to dry on the lawn. A bin bag of dried grass will last a season. Other fuels are rotten wood, papier mache made from newspaper, or corrugated paper. With all of these fuels you should be careful that they do not produce hot smoke which will annoy the bees rather than subduing them. Hessian sacking makes very good smoker fuel, but is not as plentiful as it used to be and care must be taken that the sacking has not previously been used to contain something that will be deleterious to the bees, such as dressed cereal or have been treated with flame retardant. Normally I use a small ball of lighted newspaper to start the smoker off. Once this is going well the main fuel is fed in until it too is alight and then more is put in until the smoker is tightly stuffed with fuel. This requires a little practice as it is easy to put out the fire if the smoker is stuffed too quickly. However a tightly stuffed smoker will remain alight longer and burn more evenly than one which is only lightly filled. Some beekeepers carry a small jar of rotten wood, or bits of absorbent material, soaked in methylated spirit to start the smoker.

Opening the colony

Once you have dressed and got the smoker going well you can pull on your gloves, make sure you have your hive tool and move quietly into the apiary. Gently is the operative word. Don't go clumping about the apiary causing a lot of vibration, learn to work quietly. Puff a little smoke into the entrance of the hive. Just let it waft in. The bees only need to smell the smoke, there is no need to blast the smoke in as though you are trying to kipper them. When bees smell smoke they go to the stores and fill themselves up with honey which makes them more easier to handle, but this process takes a little time so do not be in too much hurry to start opening the hive. Wait for a couple of minutes giving an occasional small puff of smoke at the entrance. Then, standing behind the hive, quietly remove the roof and place it top down on the ground behind the hive. We will assume that the hive has a super over an excluder above the brood chamber. The next thing is to push the hive tool in between the bottom of the super and above the excluder at the left hand corner of the hive and gently lever the super upwards about a half an inch. As you do this you should have the smoker in your left hand (I am assuming you are right-handed so if not reverse all the directions) so that you can puff a little smoke into the crack as you lever the boxes apart. The super is now lowered and the hive tool inserted on the right hand side and the super prised up again, using smoke as before, but lifting it higher. Put the smoker down holding the super and with both hands twist it to free it from the excluder. You must ensure that you are not lifting the excluder at the same time. Lift the super off and place it on the upturned roof behind the hive. It is always best to have the removed boxes behind the colony because on warm days in summer drones may suddenly return home and if the super is in front of the hive the drones will enter it and will still be in it when it is replaced on the hive above the excluder. As drones cannot pass through the excluder they will be trapped above it and will eventually die, blocking off part of the excluder and thus reducing bee passage and ventilation. Having removed the super immediately puff some smoke across the top of the excluder. Do not smoke heavily as even workers cannot pass through an excluder quickly and will be irritated by heavy smoking. Lever up the corners of the excluder and, if the excluder is framed, twist it loose and lift it off puffing some smoke under it as you do so. If you are using an unframed zinc excluder peel it off holding one corner. This does less damage to the excluder than holding it by two adjacent corners and peeling. Check that the queen is not on the excluder. A gentle bump on the excluder will drop her back safely into the hive. You are now down to the top bars of the brood frames and should now drift smoke across the top until all the bees have moved down between the frames. If you watch them a moment you will see the bees gradually come back onto the top bars. Those that walk around are no problem but there will be some sitting in between the frames watching you. You can see their heads moving as you do. These are the ones that may cause trouble but another puff of smoke across the top will send them out of sight. With a colony of reasonably tempered bees, as you should have, a couple of puffs of smoke in this way should have the colony completely under control and further smoke will only be needed to move bees out of the way while you are examining. You are now in the vision of the bees, whose eyes are particularly equipped to see movement. Therefore make all your movements slow and steady, without jerks or rapid hand and arm movements, and bees will take little notice of you.

Removing frames

The next thing to do is to take out the dummy frame and then the first frame of comb. This should be done gently and carefully. A few extra moments spent being careful with the dummy and first frame will always be worthwhile. I see more trouble caused by rough handling, and particularly of the first frames, than anything else. The dummy should be loosened with the hook of the hive tool on the left hand side first and then the right. This way avoids moving the hand twice across the top of the bees and alerting them. The dummy is then drawn towards you into the small empty space and withdrawn. The dummy or frames should be grasped firmly by the top bar, using the thumb and first two fingers just inside the side bars. As soon as the top bar is raised an inch or two the third and little finger should be curled under the lug on the outside of the side bar. Lift the frame upwards, taking care not to rub the bees between the faces of the one you are lifting and the next frame or against the outside wall. Learn to remove the frame without any sideways movement which will crush bees between the faces of the side bars and the hive wall. Crushing bees releases an alarm pheromone which will alert some of the other bees and make them more inclined to sting. The dummy and the first frame of comb that have been removed can, for most of the year, be placed at the entrance of the hive, leaning against the stand. If, however, your manipulation is going to take a long time, or is in a period when robbing is likely, then the dummy and first frame can be placed in a nuc hive and covered, or at worst, put back into the hive, although the latter loses the extra space and makes manipulation more difficult. Removing the rest of the frames for examination, one at a time, is much easier as the empty space of the first frame gives more room. The examination of the second comb completed the frame is carefully lowered into the brood chamber and placed in the empty space away from the rest of the frames. The space is now between the second and third frame. The procedure is repeated with the rest of the frames, care being taken that as each frame is returned to the hive it is placed with its spacers hard against the previously examined frame. On no account allow spaces to be left between the Hoffman side bars or when you have finished you may crush bees between the frame sides. One day one of these bees will be the queen and that will be a major disaster and you will not know of her loss until the next routine inspection. During examination the smoker should be kept alight and should be hooked onto the side of the hive within easy reach. You will often need to use small puffs of smoke to clear bees away from the place where you are using the hive tool, or placing your fingers to pick up the frame, so that bees are not injured or crushed. Once you have finished the examination, if you are using Hoffman frames, the hive tool can be placed in between the outside wall and the second frame and the whole batch of frames eased across the brood chamber into their normal position and the first frame and dummy replaced. The queen excluder is now replaced. If you are using an unframed slotted metal excluder the slots should be at right angle to the top bars of the frames. If they are put on the other way it greatly reduces the amount of space for the bees to get through. The super may now be returned to its former place on top of the excluder. Because the bees often cluster on the underside of the frames of the super, before putting the super back on it is worth puffing some smoke under it while it is still on the roof to make them move up out of the way between the combs. With the super back on the hive, if you wish to look in it, the crown board can be levered up with the hive tool, the examination of the super completed, the crown board replaced and the roof put on. Having finished the examination be careful to put the smoker out to prevent fire risk. The easiest way to extinguish the smoker is to stuff the end of the nozzle with grass and lay it on its side.

Examining combs

Handling Frames

Frames that have been lifted out of the brood chamber for examination should not be held horizontally, with the comb face parallel with the ground, or the comb may break loose and drop out. This is particularly so when the weather is hot, the combs heavy with brood or new comb which has not been toughened by use. Unwired comb is especially vulnerable. To examine the second side of the comb the frame should be turned through 90 degrees so that the top bar is in a vertical position and the bottom bar to the left. The frame is now turned through 180 degrees so that the bottom bar is to the right of the top bar, and then turned to bring the top bar horizontal with the bottom bar above it. The procedure is reversed to bring the frame back to the right position for return to the hive. Using this method prevents damage to the comb.

Turning a frame to inspect the second side.

What to look for

When first starting beekeeping it is necessary to spend time looking at the combs in the colony and learning all those things which will, in the future, allow you to automatically read the colony and to manage it well. The beginner should look carefully at the combs, learning the appearance of the bees on the surface and of the contents of the cells. The three types of honey bee should be found and their shape and colour noted. Note the distribution of the workers on the comb and their reaction to exposure to light and being kept out of the hive. This will vary with different strains of honey bee so do not be surprised if you see bees doing slightly different things in other beekeepers' hives: the distribution of stores, eggs, brood and fresh nectar in the cells, the shape and colour of cappings on both brood and honey. Once you have become familiar with all these things then anything unusual should be immediately noticed and enquired into. This is the basis of good management, so let us look at it in more detail.

Comb

When comb is first drawn out from foundation it is the natural colour of the foundation used. New comb is therefore usually a pale yellow which gradually turns to brown and then to nearly black as the cells are polished with propolis ready to house many generations of brood and many little feet walk over it. Foundation has the imprint of the worker cell so when drawn out it will be constructed of all worker cells. Soon the bees will alter some of the cells, most usually in the bottom two corners, to drone size so that they can produce the drones their instincts tell them they must have.

Honey

Stored honey is always placed above the brood. Fresh nectar may be placed in empty cells amongst or below the brood but will be converted to honey and put in its proper place very quickly. Notice that when honey is stored in the cells these are lengthened. This means the distance between two adjacent comb faces in the honey area is only sufficient to allow one bee to work. In the brood area with thinner comb two bees can work back to back and not get in each other's way. As the honey is put into cells it can be seen by the beekeeper as a shiny liquid. As soon as the cell is full the bees seal it in with a wax capping. The capping on honey will be seen to be different from that over brood. The cappings over brood are rougher and made so the pupa inside can breathe. The cappings over honey are smoother and impervious. It is often difficult to see the hexagonal top of the cell and the cappings are often wrinkled. The capping of honey always starts off white, or pale yellow, as the bees use freshly made wax but in time becomes darkened by travel staining, again the results of many little feet walking over it.

Pollen

The loads of pollen brought in by the foragers are placed in cells around, and sometimes mixed in with the brood. House bees then come and push it down in the cell. Bees requiring pollen eat it from the cell and at this time the pollen will have a matt surface. If for some reason pollen is not required, i.e. a swarm has been lost and there is no young brood in the hive, then the bees cover the surface of the pollen with honey to prevent it from growing mould and the surface of the pollen is then shiny. Pollen from different flowers is a different colour and can vary from white to black, with yellows, reds, blues and greens in between.

A pollen load being collected on the hind legs of the bee for transport home.

Left: A frame containing a frame of many varieties of pollen.

Right: Eggs and young larvae, some being fed by workers.

Eggs

Before the queen lays in the cells they are polished in preparation by the worker bees. The polish is so good that a small point of light is reflected from the concave surface of the cell base and it is quite easy to mistake this reflection for the presence of an egg, particularly in poor light conditions. Eggs are laid singly, glued to the base of the cell in a horizontal position that is parallel to the walls of the cell. The queen lays methodically, working spirally over the face of the comb. This will lead to neighbouring eggs, and later brood, being of the same age. If there are more than one egg in the cells then there is usually a problem. However, new queens sometimes lay more than one egg in a cell, whilst they are 'learning'. This behaviour lasts a short time

Larvae

Brood in unsealed cells is often termed 'open brood'. At this stage the larvae can be seen in the cells. They should be pearly white and curled up on the base of the cell. When just hatched they will be very tiny, no longer than the egg, and will be floating in a whitish translucent fluid, brood food. Often at the beginning it is easier to see the food than the larva itself. However, this soon alters as the larvae grow very quickly and become very easy to see. They should still be neatly curled at the bottom of the cell and pearly white in colour. Any variation from this appearance is a sign of trouble – see starvation, chilling and disease Chapter 7. In these larger larvae it is often possible to see a coloured line running down their bodies. This is normal. It is the coloured pollen which they have eaten, filling the gut and visible through the body wall. By about six days after hatching the larva is full size and fills the base of the cell. It is still curled up but it is a very tight fit.

Sealed brood

As soon as the larvae are full sized the worker bees seal, or cap, the cell and the larvae can no longer be seen without removing the cappings. On worker larvae the workers seal the cells with a small, low, domed capping while drone brood is sealed with a high dome, three or four times the height of the worker capping. As workers do not add fresh wax to make the capping the colour of the cappings will match the colour of the comb they are on. Old comb will have dark brown cappings, new comb pale yellow cappings. All the cappings on a single comb should be the same colour. If single, or small patches of, cappings are darker than the rest this could be trouble – see disease.

Queen cells

For most of the year there are no queen cells in a colony. There is, however, during most of the active season the precursors of queen cells normally called queen cups, see overleaf. These cups are constructed around the edges of, and in holes in the comb. Looking inside the cup in the early season you will find the interior surface quite matt and none reflective. Later the bees polish the inside and the surface reflects light. The cups are then ready for the queen to lay in them. Once the queen has laid an egg in the cup and conditions are right in the colony for them to make new queens (otherwise they eat the egg), the bees extend the walls of the cup turning it into a queen cell, feed the larva heavily with royal jelly and cap the cell on the eighth day (from when the egg was laid) when the larva is full size. The colour of queen cells again matches the colour of the comb on which they are built.

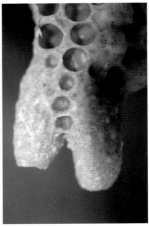

Queen cells ready for sealing.

The queen has emerged leaving the hinged lid.

Workers

Most of the bees you will be looking at on the comb will be workers. They can range from almost black to brown with two or three amber-yellow stripes, have a pointed rear end, fairly short legs and wings that almost cover the body.

Drones and workers on a comb containing pollen.

Queen

A good queen will be about twice the length of her workers and considerably thicker in build. Colour can vary from almost black to plain dull yellow or sometimes with alternate broad light and dark brown or amber stripes on the abdomen. Her abdomen is pointed, the thorax is bald, her legs long and rather spidery in appearance and her wings only reaching about half way back along her body. A good well-produced queen usually stands out above her workers on the comb and is easily seen. On the other hand emergency queens can be little larger than workers, are hard to find and useless.

Drones

By the end of April in most years there are plenty of drones about in the colonies and they will be there until the end of August. They are much heavier in build than the two female castes, not as long as the queen but as broad or broader. The end of the abdomen is quite blunt, the thorax is large and almost square and hairy, while the head is large, round, shiny black because the very large eyes almost reduce the 'face' to nothing. The very large wings cover the body.

4. The Bees have arrived

The arrivals

The bees can have arrived as a nucleus, a package or you can have collected a swarm. We will deal with the reception of each of these in turn, and in the above order.

The nucleus

Ask your supplier when they are likely to have the bees ready for sale, and ask them to notify you how they are being sent at the time they dispatch the bees. This will allow you to meet the nucleus, ensuring it does not get delayed in transit, and to get the apiary laid out ready to receive the bees. Usually the nucleus arrives in a ventilated box and has been on the way for at least a day. As soon as the nucleus arrives, it will probably be buzzing. Without opening the nucleus box, pour a small cup full of water through the top screen onto the bees and frames. This will make them happy as often they may be getting quite thirsty. If you cannot put the nuc out right away put it in a cool dark place and if they start to buzz loudly again pour in another cup of water. As soon as possible take them to their new home, stand the nuc box on the floor of the hive in the exact position they are going to take up and open the entrance to the nuc box to allow them to fly. This is best done in the evening. This must be done in the exact position the colony is to occupy, because they will learn the landmarks for this position as soon as they fly out. After this if you move the hive some bees may be lost, and certainly the flying bees will be greatly inconvenienced. See 'Moving Colonies' page 110. Once the little colony is opened cover the top ventilation and place something over the top of the box to shield it from the sun's heat, or from rain. The roof of the new hive would be ideal. Put a brick on it to prevent it being blown off. The bees should be allowed to settle down for an hour or two before being transferred into their new hive.

Transferring

This is the technique of putting the honey bees and their combs into a new hive, in this case the nuc from its travelling box to its permanent home. The weather should be dry and warm (over 60°F/30°C). Dress ready to handle the bees and light the smoker. Get everything ready and to hand by the nuc. You will need a floor, entrance block, brood chamber, crown board and roof, four frames of the correct size with wax foundation ready inserted, and I would advise the use of two dummy boards. Smoke the nuc and move it a couple of feet to one side. The floor will be already in place on the stand. Put the brood chamber on it with the entrance block in place. Gently smoke the nuc once more and then remove the top cover of the travelling box. A gentle puff of smoke again. Remove the frames one at a time and place them in the same position, in the same order, in the new brood chamber. While doing this you can have a quick look to see if there are eggs and brood present. You might also check to see if you can catch sight of the queen. Do not spend a long time looking for the queen, although it is nice to know the queen has gone safely into her new home. Once you are experienced you will always make sure the queen is transferred safely. Having got all the frames safely into the new brood chamber, add the four frames of foundation. Some beekeepers would put two on each side of the nuc, but I prefer to place them all together on one side because this allows comb building to occur in one place rather than two, and it is easier to assess the expansion of the little colony if this all occurs in one place. Beekeepers with two dummies should put one on each side of the eight frames, and push the whole to the centre of the brood chamber. If there are bees left in the travelling box shake these out into the hive or, if you find this difficult, prop the travelling box up into a position where the side of the box is resting against the stand just below the entrance of the hive. The bees will then crawl up to join their friends in the hive once the colony's odour reaches them. The Miller feeder is now put on the hive and about a gallon (4.5 litres) of syrup given to the colony to help it in drawing out the foundation. Do not forget to pour a dribble of syrup down the bee entrance of the feeder onto the top bars of the frames as this lets the bees know that the syrup is there. Colonies do at times fail to find syrup in a feeder, especially if the nights are cold, unless they are made aware of it. The syrup used should be 'heavy syrup' that is made of 2 pounds of white granulated sugar to every 1 pint of water. A gallon of syrup will therefore be made of 8lbs sugar in four 4 pints water (1¾ Kg sugar in 1 litre water). The crown board is then placed on top of the feeder. Block the two escape holes with flat piece of wood or slate, and put the roof on.

Package bees

Dealing with a package is slightly different because there are no frames with the bees. The hive should be got ready as before but in this case you will need at least six frames of foundation ready to put into the hive. Make arrangements with the supplier to let you know when the package will arrive so that you can get everything ready and meet it. The bees come in a box with two long sides made of wire mesh. The top of the box has a hole in it through which a small tin of syrup, or block of candy, is placed so that the bees can feed from it while on the journey. The queen is confined in a cage and fastened into the top of the box. She may be a young mated queen which has had no previous contact with the bees in the package. The fuss of the journey will, however, usually mean she will be accepted by the bees when released. When the package arrives I would put it, for an hour, in a cool place and pour some thin syrup (1 lb sugar to 1 pint water) through the mesh onto the bees. This is particularly necessary if the package has been supplied for the journey with candy, not syrup. Even if they have been supplied with syrup they will only be using it for their immediate requirements. Pouring some over them will cause them to clean it up and thus fill themselves, making them more contented and amenable to handling. The next job is to put the bees into the hive. This is best done in early evening. The hive floor and brood chamber should be placed on its permanent site and the brood chamber half filled with frames of foundation. The syrup container, or candy block, and queen cage should be removed from the box and put safely to one side. Space out the frames equally across the brood chamber they will then be about 1½ inches (3.8cm) apart, and pour the bees from the package through the top hole of the travelling box into the hive over the frames. The bees are usually quite docile after their journey. Get as many of the bees out of the box and into the hive as you can and then, if the evening is warm place the box just below the entrance so that the rest of them can crawl directly from the box through the entrance into the hive. If the weather is wet or cold pull off the mesh from one side of the box and empty them all into the hive. Carefully push the frames up tight onto their spacers, making sure that you do not crush bees, and put a dummy on each side of the frames. Separate the two centre frames and place the queen cage between their top bars and jam it tight so that it will not slip down. The queen cage should have the protection over the candy hole removed and a nail pushed through the candy to make a hole, about ⅛ inch (3mm) in diameter, through its centre thus helping the bees to eat their way in and free the queen into the hive. If the candy is hard most of it can be scraped away to prevent the bees being delayed. Stuff a sack or rolled up newspaper into the space beside the frames and the wall of the brood chamber, to prevent the bees setting up a cluster in the space, close the hive and feed as detailed for a nucleus. The whole brood chamber could be filled with frames of foundation, but the package will settle down and build up faster if they are condensed on fewer frames at the start. An established beekeeper who has a reserve of drawn comb would give this to a package of bees as they develop much faster if put straight away on drawn comb.

Swarm

Here we will deal with the technique of putting a swarm into a hive. How to collect a swarm is dealt with on page 74. A swarm is, by instinct, in an optimum condition to build comb. The beekeeper should always take advantage of this by giving them foundation to draw out. Of necessity the beginner will have to use foundation as they will have no drawn comb.

Hiving a swarm – traditional method

Swarms are best hived in the evening

There are two ways of hiving a swarm: the traditional way and the quick way. I would recommend the former to a beginner as it is a very satisfying and absorbing method.

The hive is placed on its permanent site, floor, entrance block, brood chamber containing eleven frames of foundation, feeder, crown board and roof. A piece of board, hardboard or chipboard is placed sloping down from the entrance to the ground. If hardboard is used it should be rough side uppermost and must be supported at the centre, because a swarm will weigh some four or five pounds and may cause it to collapse. The swarm is emptied from the box or skep in which it has been collected onto the sloping board as near to the top as convenient. The bees will spread out in all directions until the ones at the top find the hive, and possibly smell the wax. These bees will start to fan their wings and to emit a rallying scent from the Nasenov gland. As the other bees smell this scent they will turn and move towards it and the whole swarm will gradually move into the hive. Complete entry of all the bees can take several hours. Feed at least a gallon of heavy syrup a couple of days later when all the honey the bees have brought with them has been used to draw the foundation.

Cloth covering the sloping board.

Foundation in place, before swarm is shaken onto the cloth.

Bees shaken onto cloth.

Spotting the queen as the bees march into the hive.

Hiving a swarm – quick method

The floor and empty brood chamber are set up in their permanent position on the stand. The frames with foundation, the feeder and syrup and the rest of the hive are close to hand. The entrance to the hive is completely closed. Use the entrance block and stuff the small entrance with grass, paper or cloth. The swarm is then emptied into the brood chamber and the frames of foundation are gently lowered onto the pile of bees on the hive floor. Do not push the frames down or you will kill bees, just lay them gently on the top of the cluster which will then gradually crawl up the foundation. As the cluster transfers itself onto the foundation the frames will sink into the box and can be adjusted onto the runners into their normal position. As soon as the frames are in position and squeezed up tight, with a dummy on each side if you have them, the feeder and syrup can be put on and the roof placed in position on top. The entrance to the hive should now be opened by removing the temporary obstruction. The reason for closing the hive entrance while the swarm is shook in is because queens of swarms will sometimes walk out of an open entrance and fly off, the swarm will of course soon follow her and may be lost. Feed a couple of days later.

Observations

The good beekeeper should always be an observant person. Those of us who are not have to train ourselves to observe everything that happens around and in the colony. So start right away to look at the bees after they have settled in the hive and see what is going on. If the weather is warm, and not raining, then the bees of the colony should be flying within a few minutes of transferring. If there is bee forage within a few hundred metres then it will not be long before some of them will be returning with pollen on their legs. Certainly they should be by the next day and the entry of plenty of pollen, not just the odd load, will indicate that all is well with the colony.

You may find that a few, ten or a dozen, dead bodies have been thrown out during the night, but do not worry about this, it is normal. Some individuals die every day and a few may have been injured during transferring. If a lot, 50 to 100 die something is amiss and you need to think if you have done anything wrong such as painting the hive with a substance containing insecticide, or even placing the bees near an insecticidal strip or spray, before hiving. Insecticide fly papers have been known to kill bees forty feet away in another room. In the latter case there should be no further deaths and the little colony should survive. In the former case the deaths will continue until you remove the bees from the hive and transfer them to a new untreated one. They could go back into the travelling box in the case of the nucleus. Don't be worried as this very rarely happens. Usually the little colony is fine and begins to grow right away.

Have a look in the feeder to see if the bees are taking the syrup. If they are not pour some more syrup down the bee entrance to the feeder to encourage them. Usually there is no difficulty and the feeder is emptied in short time and can be removed. When removing the feeder put your veil on and light, and use the smoker. This may not be necessary when you become experienced but is advisable for a beginner. If you have a glass quilt then this is ideal. Put it on the hive when you remove the feeder. You can then go down most mornings and quietly take off the roof and look at the bees through the glass. You will thus be able to see what is going on without needing to dress up or light the smoker. You will be able to watch the expansion of the cluster of bees onto the new frames as they draw out the foundation, and probably be amazed at the rate they do this.

Watching the nucleus

Having put the nucleus into its hive you will be able to watch the colony flying and bringing in different coloured loads of pollen. Because you have fed the colony some syrup they will be fussing about the entrance and flying quite strongly if the weather is nice, or just one or two flying off for short distances if the weather is inclement. If they are not fussing around the entrance look in the feeder to see if they are taking syrup. If they are not lift the board or glass strip covering the bee access point and slop some of the syrup down onto the frames so that there is a trail to lead the bees into the feeder. Usually the bees will be feeding well and you can leave them alone until they have removed all the syrup from the feeder. The house bees will have taken it down and stored it in the cells of the comb. When a week has past, and the feeder is empty, it will be time to open the colony and examine it once more.

Examining the nucleus

When you open colonies you should have in your mind exactly what you are going to look for. In the present case you need to know:

1. Is the queen alright, and has she has she survived introduction to the nucleus?

2. Is the nucleus expanding and building out the foundation you have given them?

3. Has the little colony sufficient stores to survive on until you examine them again?

Put on your beekeeping clothes and light the smoker. Take with you enough frames fitted with foundation to fill the broad chamber. This will usually be three frames. Walk gently down to the hive and smoke the entrance. Take the roof off and place it on the ground top downwards. If there is a Miller feeder remove the crown board and lift the glass strip and puff a little smoke under it, then take the empty feeder off, puffing a little smoke over the top bars as you do so. Place the feeder in front of the hive in a position which allows any bees in it to crawl up to the hive entrance. Puff a little smoke over the top bars to send the bees down. Hook the smoker on the side of the hive and, working from the side away from the original frames of comb, that is those which were part of the original nucleus, remove the dummy and then look down at the frame. There is no need to lift it out. You will be able to see if the bees have started to build cells on the foundation. If they have not started, which is most likely move the frame into the gap created when removing the dummy board. Now repeat the process with the next frame. A good nucleus will probably have started to draw out the cells on this comb in which case, if they are working on the cells of the comb, lift it out for examination. Notice how they usually start pulling out the cells at the centre of the frame first and gradually work outwards, so that the walls of the cells are highest at the centre of the foundation and gradually get lower as one moves away in any direction until finally reaching foundation which has not been touched. How much they have done will depend upon the strength of the colony and the weather and forage conditions reigning at the time. If the bees have started on frame 2 then frame 3 should be lifted out and examined. Look towards the centre of the comb, into the cells to see if the queen has laid any eggs in the cells. If she has not there will probably some cells at the centre containing stored pollen and stored syrup. The 4th frame, the one nearest to the original nucleus, should have its comb completely drawn out and you should be able to see eggs and young larvae in the cells. As you go through the frames note the amount of sealed stores present. The colony should have the equivalent of about one complete frame of stores. If it has less than this, give them a further gallon of syrup. It is likely that this will be necessary because the bees have to draw foundation. A Modified Commercial, 16 by 10, frame would hold about 6½ lbs of sealed stores, and a British Standard deep frame holds about 5 lbs.

Building up the nucleus

Examination of the colony is finished. Add the rest of the frames of foundation to fill the brood chamber, taking out one of the dummies and leaving the other on the foundation side of the frames. Close down by putting on the crown board, or glass quilt, put on the roof and make out your record. It is a very good plan to keep simple records to aid your memory, and allow you to assess your beekeeping at the end of the year.

Continue to feed the bees until most of the frames of foundation are drawn out. Usually another two gallons of syrup will suffice.

Spreading brood

This is a technique that will speed up the development of a colony. The technique needs a little experience so as not to stretch the colony beyond its capabilities and result in chilled brood. See page 118.

Problems

1. The bees are not taking syrup from the feeder. If Miller feeders are painted and finished with a smooth shiny surface, bees cannot, or will not, walk over the surface to get at the syrup. Aluminium feeders may also be too smooth, or have lost the perforated metal inserts in the entry ways. The cure in both cases is to roughen the pathways. Perforated metal inserts is the usual way. Try this for a day but if the bees still ignore, or only a few enter the feeder then take it off and look at the frames of foundation. If this is being drawn out then the problem is almost certainly a feeder problem or the weather is very cold and the bees will not venture away from the cluster. The latter is unusual because by the time nuclei are sent out the weather is reasonably warm. So have another go at roughening up the walk ways in the feeder.

2. There has been little or no work done on the foundation, it is almost the same as when it was put in the hive. This is a very unlikely condition and it probably means the queen was lost during transport or when transferring from nuc box to hive. Looking at the combs which came with the nuc there are no eggs and only old larvae in the open brood, and they should be making one or two queen cells. These queen cells will result in a very poor queen taking over the colony. A new queen will be required, so it is best to buy a good queen and introduce her into the colony. For introduction methods see page 115.

Package bees

What to look for

Package bees are much more likely to have problems than are nuclei. They should really be hived onto drawn comb containing stores of honey and pollen but this is impossible for the beginner who only has foundation to put them on. Watch for bees flying, fussing about at the entrance and bringing back loads of pollen to the hive. If plenty of loads of pollen are being carried in, then probably all is well.

Examination of colony

Method and purpose of examination will be the same as for the nucleus. The queen cage should removed, and examined to making sure it is empty. Examine the frames quickly moving directly to the frames in which the foundation is being drawn out. Take them out of the hive, one at a time, and look for eggs and young larvae. Once these are found you know that the queen has been accepted and is laying. Make sure there are sufficient stores, honey or syrup, in the colony. Feed heavy syrup if there are any doubts. If the colony is drawing out the foundation give it more frames of foundation so that it can continue the good work. I would suggest giving the same number as the colony has drawn out. That is all one can do at present so close up the hive.

Building up package bees

It must be realised that in this case, because there was no brood in the original package, the worker population will decrease for the first 21 days until the first brood has had time to go through its development period and emerge. After this period the population should gradually build up to full size, but this will take some time. During this early period therefore one can do little to aid the colony's expansion until it has had time to establish a brood nest of several frames. It is best for the beginner to let the colony build up at its own rate and not to try to hurry things. Making sure that the colony never runs short of stores is all that needs to be done.

Swarm

The hived swarm is in much the same position as the package bee colony but has one great advantage in that the bees have an instinctive desire to work very hard to re-establish themselves in their new home. If it is a prime swarm, that is the first swarm out of its colony that year, it will have with it the old queen. Casts or afterswarms are further swarms following the departure of the prime swarm usually with a newly emerged queen. Afterswarms contain virgin queens, which have not yet mated. The virgin queen may take a while to mate before she can lay and eggs appear in the cells. However, if the little colony draws out the foundation in readiness and polishes the base of the cells then almost certainly all is well and they have a queen. Very small casts may contain more than one virgin queen and will usually refuse to settle in a hive and will swarm out again the same or next day. The beginner had best let them go or turn them over to an experienced beekeeper to deal with.

Building up swarms

The swarm with a laying queen will have new workers emerging in about three weeks. Meanwhile the older workers arriving with the swarm will be dying off and the colony will shrink in size before increasing again as the new brood emerges. They do however have an urge to work and build up rapidly if they have enough food. Always ensure that the little colony has a surplus of stores and should always feed heavy syrup if there is any doubt whether they have enough for their needs.

General

Once these colonies, whatever their origin has six frames well covered with brood (at least three quarter of the area of each frame) then the first super should be put on. The colony will then be managed in the normal way as described in the next chapter.

5. Colony Management

General

The small colony will quickly grow larger. Do not delude yourself that a nucleus or a prime swarm will not get a crop in the first year, or even swarm if given the chance to do so by inadequate management. What follows deals with the routine management of any colony throughout the year. The aim is to ensure that everything goes well with the colony, to iron out its problems and to prevent it swarming. To do this it should be examined at weekly intervals and given the attention needed. The interval between inspections can be up to ten days but to many people weekly inspections are most convenient, as they can then look at the bees during the weekend.

Inspections

When?

Inspections can be carried out at any time when the weather is fine and the temperature is over 50°F / 10°C. If the colony is flying strongly then it will always be alright. It is possible to inspect colonies at lower temperatures if one is very quick, but this comes with experience; beginners should stick to the above temperature. They should try to learn to do their examination of the colony, and the jobs that they decide require doing, as quickly and efficiently as possible. Once they have gained proficiency in handling there is little need to have a colony open for more than 10 to 15 minutes. Longer periods tend to disrupt the colony unnecessarily and to increase the likelihood of angry bees starting to sting.

What for?

Routine inspections of this sort should always be carried out to a set plan of campaign so that no essential points are missed. The routine questions to be asked are:

    1. Are there eggs present in the brood nest?
    2. a. Is the colony building up in size?
        b. Are there queen cells in the brood nest?

3. Has the colony sufficient stores to last until next inspection?
4. Has the colony sufficient room for its future needs?
5. Any abnormality or disease to be seen?

These five questions should be asked every time the beekeeper examines a colony.

Recording

It is best to keep records of your inspections to help you the next time you open the hive. What is wanted is not a long essay but a short series of coded messages notes which can be taken in at a glance. Essential details are: origin and age of queen and whether she is clipped and marked, the 5 questions answered, and make notes about temper, amount of honey taken and amount of syrup fed.

Record card: Queen – Clipped/marked   **Colony No.**   **Apiary name**

Date of inspection	Queen eggs	Col.b/u Q.cell	Stores	Room/ supers	Diseases	Remarks

An example of a simple record card, which can be stored under the roof in a plastic pocket. Everyone makes up their own shorthand system which they can remember easily.

First question

To find eggs in quantity when manipulating a colony is extremely important because it decides whether the queen is still laying. The answer should be obtained before any management manoeuvre is carried out. Failure to follow this practice may, by error, allow considerable damage to be done to the colony. Even if you see a queen you must still ensure eggs are present because she could have stopped laying, it could be a young unmated queen usurping the colony by supersedure or one taking over after swarming which may or may not be mated. Observation of the number and distribution of the eggs in a colony is one of the factors used in assessing the value of the queen and therefore of the colony.

Assessing quality of queen

The quality of the queen is of paramount importance to the value of the colony as a whole. One could say she is 'the colony' because she carries within herself all the inherited characteristics which will appear and once she has been replaced the colony will be a different entity with different characteristics. The queen's quality is an expression of her origins and comes from three main directions: the characteristics she inherited from her parents, the quality of the care and nourishment she was given during her larval period and the characteristics the workers express gleaned from the drones with which she mated. The queen's quality is assessed from two directions: her ability as an egg-layer and the performance of her colony.

Egg laying ability

A good queen lays in a very systematic way, working in a spiral fashion and filling adjacent cells very quickly. When brood emerges house bees very rapidly clean and polish the vacated cells and get them ready for the queen to lay in again, which she does within the next few hours. Therefore when examining a colony during spring and summer look into the open cells in the brood area to see what is in them. If they are polished and have not been re-laid, although you have already found eggs elsewhere, this should cause you to look more carefully as there may be something wrong. If there are a lot of cells in this state then either the queen is beginning to fail or the colony is making queen cells. Queens can fail and become incapable of maintaining a full sized colony, at any age. Failing to re-lay in cells from which brood has hatched is one of the first signs of failure. A failing queen should be replaced with a good young queen, see page 116. In the height of the season the reduction or absence of eggs, could mean that the colony is making queen cells and preparing to swarm. The workers reduce the queen's food intake, causing her to slim down so she is able to fly. A good queen in full lay cannot get airborne. In this case the colony should be dealt with to prevent the loss of a swarm. See 'Swarm prevention' page 66 and 'Swarm control' page 69. At the end of the summer, during or at the end of the final nectar flow, the queen will reduce her egg laying rate and the brood area will naturally diminish in preparation for winter. This is not a cause for anxiety but perfectly normal.

Brood pattern

The systematic way in which the queen lays eggs and the viability of these eggs is reflected in the pattern of open and sealed brood. For instance a good queen laying in a 16 x 10 brood frame, and filling about three quarter of the cells, will put eggs in about 5,000 cells taking about 48 hours to do so. The laying will spread out, concentrically, from the centre of the comb and therefore after five days, when the last of the eggs have hatched, it will be seen that the oldest and largest larvae are in the centre and they get smaller towards the periphery of the area. In another four days the central cells will be being capped over and capping will continue to spread outwards for a couple of days until the whole of the brood area on the comb is sealed. Twelve days later the bees will be emerging at the centre of the comb and by next day the empty cells should be polished and re-laid. This is the series of events one would see in a colony with a young high quality queen. The brood pattern of the majority of queens will show some variation from this ideal. How far the deterioration has progressed will indicate whether the queen is still good enough to head the colony or whether she should be replaced.

How does one recognise the deterioration?

Looking at a comb of sealed brood there will always be some open cells in view. Some of these will contain stored pollen which is normal, but others will be empty or more likely, contain an egg or a young larva. Probably what has happened is that some of the eggs have failed to hatch, they were not viable, the bees have cleaned the cells and the queen has been around and re-laid them. Not more than 5% open cells in sealed brood is acceptable, that is five open cells in each area of 2 inches by 2 inches (5 by 5cm). When more than this occurs it should be looked into as it may indicate the presence of disease, see page 76 et seq, or it may be the queen is beginning to fail. As the queen's failure becomes worse her laying becomes more haphazard and the mixture of larval ages in an area of comb increases. Ultimately the brood pattern will deteriorate into eggs, small and large larvae and sealed brood all present in quite a small area of comb. The queen should be replaced before it gets to this state. For replacement of queens and queen introduction see page 115 et seq.

Second question

The second question is in two parts. Question 'a' is used in the early part of the active season while the colony is still building up its population after winter, while 'b' is used once the colony has reached a size where it needs supering and treating for swarm prevention. The change from 'a' to 'b' will occur when the colony has reached six or more frames in which the brood covers at least three quarters of the comb area on each side. Note the number of combs containing brood. In the early season the beekeeper should expect to see expansion of the brood nest at each inspection. The expansion can occur in two ways, by an increase of the area of brood on each comb yet the number of frames with brood in the comb remaining the same (this is usually the expansion made by a small and possibly poor colony) or the expansion may be shown by the presence of a greater number of frames with brood, the queen having spread her laying out onto one or more fresh combs. The rate at which expansion occurs depends upon the quality of the queen, the initial size of the colony, the weather and the amount of endemic disease in the colony. It is during the building up stage that the skill of the beekeeper has great effect on the subsequent honey crop. This skill comes with experience. Once the build-up period is over then dealing with the colony becomes much easier because it is easy to see queen cells and, once they appear, the beekeeper can apply his adopted method of swarm control. See page 69.

Third question

This is just a matter of ensuring that the colony has enough stored honey to last them until you will see them at the next inspection, even if the weather is atrocious and the bees are unable to go out of the hive for the whole period between the inspections. The colony should never be left with less than ten to fifteen pounds of stores that is two or three standard frames of honey, or 200 to 300 sq inches (500–750sq mm) on each side of the combs. This will be about 1½ to 2½ inches (38 to 64mm) depth across the top of each of eleven frames. Once there are supers on the colony the question about stores can often be answered by the weight of the supers as they are lifted off.

Fourth question

Once colonies are full sized and have a complete complement of frames, this question is really asking do they need supering. Supers, or shallow boxes, are put on the hives for two reasons, one to hold the honey stores and two to accommodate the number of workers as the population builds up towards a peak. If the workers are congested on the comb they will be more likely to start making queen cells in preparation for swarming than if they have plenty of comb space to spread out onto. The general rule for supering is that the bees should never be allowed to get to a state where they are using all comb available. In this context 'using' means just standing on it. There should always be comb free of bees. The first super should be put on, above a queen excluder, before the bees are right across the brood chamber and the second put on when the bees are three quarters across the first super. The second super could be put on below the first one to draw more bees out of the brood chamber. After this all further supers should be placed above the ones already on the hive, as it is then only necessary to lift the crown board to see how the bees are getting on with their job. Supers should be added as honey is stored, combs begin to fill and honey is stored as the bees begin to crowd into the supers. Remember when nectar comes in it contains a considerable amount of water which is later evaporated off by the bees. Therefore nectar requires considerably more comb space, than the finished honey. Keep extra space ahead of them until the nectar flows are coming to an end.

When does the nectar flow end?

This you will have to find out from your local beekeepers because it varies from place to place. There be may be small nectar flows at any time if the weather is suitable, but the main nectar flows in a particular area are moderately constant in their times of starting and stopping.

Drawing super foundation

When ordinary (not Manley) super frames full of wax foundation are put on the colonies for drawing the frames should not be spaced wider than 1½ inches (38mm), centre to centre. If wider spacing is used the bees may well build their own comb in between the frames and ignore the foundation. Once the combs are fully drawn for this spacing, and before the honey is sealed, the spacing can be widened to 2 inches (50mm). This will reduce the number of frames in the super to eight. The bees will draw the combs out further so that each comb will be thicker and therefore hold a greater amount of honey. This will provide economy in the use of frames and will make extracting easier by reducing the number of frames to be handled for the same amount of honey. Beekeepers using castellated runners in their supers will need to keep some with ordinary plain runners in which to get their foundation drawn Manley frames are spaced at 1 ⅝ inch (41mm), and have wider bottom bars so that the bees draw them out correctly without a change of spacing. When one has a number of super frames with drawn comb any further foundation which one wants drawn can best be put in the centre of ready drawn frames. For example a block of four frames with foundation can be put in the middle of a super, this is where the heat is greatest and the bees can most easily manipulate the wax.

Fifth question

This is difficult for beginners. No one can expect you to recognise disease; this comes only after considerable experience of beekeeping. My advice to those just starting is to get to know what is normal as soon as possible and if you see anything unusual in the hive and cannot satisfy yourself what it is from reference books then ask advice of a competent beekeeper. See Chapter 7, page 76, for details of honey bee disease.

Maintenance of equipment

In addition to the routine inspections dealt with above the colonies should be looked at with a view to keeping the hives and equipment in good order. Any damage to brood chambers or supers should be repaired as soon as possible. If you have plenty of extra equipment then you can change the damaged boxes for good ones and stack the damaged ones to be dealt with in the winter. Roofs should be inspected to ensure they are not leaking. Stands should be periodically inspected and re-levelled if necessary. Solid floors, if used, should be removed and cleaned if dirty in spring. If using open mesh floors debris should be cleared from the ground underneath. This is especially important if the collecting tray is in position for any length of time. Any debris in the tray will encourage wax moths. The apiary should be kept clear of tall weeds especially in front of the hives, where weeds will impede the bees going in and out and reduce, or make more difficult, the ventilation of the colony. Frames should be cleaned of brace comb as soon as the weather is suitable in the spring. This job can be done more easily when the weather is warm and wax is soft, not brittle. In the cause of hygiene, bits of comb should not be thrown about in the apiary but should be collected and the wax recovered.

Comb care

The quality of brood combs in a colony should be noted and any with more than 10% drone cells should be worked to the outside of the brood nest in summer and removed when the brood nest has reduced in size and the queen stopped laying eggs in them. These combs can then be melted and the wax recovered. Colonies will have a certain amount of drone comb in the hive and if you take it all away they will convert some patches of the worker cells to drone cells, usually in the bottom corners of the frames. It therefore pays to have one comb with some drone cells on the outside of the brood nest to stop the bees spoiling too many of the other combs.

To provide well drawn replacement combs it is a good idea to put a brood chamber filled with frames of foundation onto a strong colony as a super. If this can be done when there is a good nectar flow in progress the comb will be beautifully drawn. Currant advice is to replace comb on a regular basis, at least every third year.

Management

General

The management of honey bee colonies will vary with the climatic conditions and the type and density of the bee forage in the district. The beekeeper's colony is flying free and is little different to a wild colony of bees. The beekeeper can only aid it in overcoming its problems as quickly and as painlessly as possible. The honey bee is basically a tropical insect and therefore the winter period is, in many areas, its time of greatest risk and this risk increases as the winters become longer and the summers shorter. To provide the beekeeper with a crop of honey the colony has first to bring in enough nectar to provide for its immediate requirements and then to store, as honey, a surplus which has to be sufficient to provide it with the necessary food during the periods, in both summer and winter, when there is no fresh nectar coming in. What the beekeeper takes as his crop is the excess over the requirements of the colony. In areas with long winters it is possible to help by feeding the colonies sugar syrup for the winter period. In fact if the winters are long then sugar is better for the bees than honey because sugar is metabolised to carbon dioxide and water with no indigestible residue whilst honey has some indigestible substances, including lots of pollen skins, which help to fill the bowel of the bees, indeed overfill it if the winter is very long with no flying days. After the winter colonies must be built up to full size in time for the main honey flow in the district and must be prevented from swarming or the beekeeper's crop from the colony may be lost for that season. The size of the surplus, the beekeeper's crop, will depend upon the quality of the colonies, the presence of the good forage plants in the vicinity and whether the good warm weather arrives when the forage plants are in bloom.

Winter preparations

The requirements for good wintering are a young queen, a lot of young bees, freedom from disease and plenty of stores. The beekeeping year starts at the end of the summer after the main nectar flows are over and the beekeeper's crop of surplus honey has been removed. Only colonies which are adequately prepared for winter are likely to give a crop the following year. Towards the end of the active season the colonies should be watched and any queens which are of doubtful quality should be replaced. The replacements can either be bought from a specialist queen producer or the beekeeper can have produced, earlier during the summer, enough queens for the likely requirements of the colonies and have them, mated and laying, waiting in nuclei until required. In your first few years of beekeeping the endemic diseases should not be a problem but as time passes they will increase in importance and should be catered for. See Disease page 76. Colonies should be fed down for winter with at least 50 to 60 lbs of stores. This means that if there is 30 lbs of stores already in the brood chamber then they should be fed at least 24 lbs of sugar made up into heavy syrup. See Syrup, below and page 119. This will provide 28 lbs of stores and set the colony up with about the equivalent of 60 lbs of honey for winter stores. The feed should be given as fast as possible. I find a Miller type feeder the best to use. See 'Feeders' pages 28–31. Finally the beekeeper should ensure that the colony and hive are protected from pests such as mice and woodpeckers. See 'Pests of the honey bee' page 87–89. It goes without saying that the hive should be watertight and secured against winter storms, shallow roofs, 6 inches deep or less will need tying, or weighing down. Having done all the above the colonies should be left alone until the new season starts.

Start of season

In the new year the timing of the first work on the colonies will depend upon when the first heavy nectar flow is expected to occur. Traditionally colonies are not touched until April, when the temperature is usually around 50°F/10°C, and colonies can be easily manipulated without damage.

Spring assessment of stores

It is advisable to gently lift the colonies to assess their weight during the latter part of February. They should not have used more than about 10 to 12 lbs of their stores up to this time, but if any feel light they should be fed heavy syrup in a contact feeder – that is a tin or bucket feeder placed right on the top bars of the colony inside an empty brood chamber or super. Most colonies get through the winter until March after which they may starve if they were not given adequate stores the previous autumn. This will happen most easily to your best colonies as they will be increasing in size most quickly and have many larvae to feed.

Syrup

Sugar syrups made to feed to the bees should only be made with clean granulated white sugar, particularly if the syrup is to be used during the winter. Dirty, raw or brown sugar is detrimental to the bees and in some cases definitely toxic, and should not be turned into syrup. The syrup should be made in the proportions of 2 lbs sugar to 1 pint water; this has been shown to be the most effective strength. The easiest way to put sugar into a bucket and mark the level, then add warm water to that level and stir till the sugar has dissolved.

Spring work

Once the temperature rises above 10°C/50°F during the day, about the beginning of April, then beekeeping can start in earnest and colonies can be examined to see how they have come through the winter. Colonies will come through the winter with considerable variation in size and therefore the first job of management in the year is to work out what is wrong with the small ones and to rectify their problems. The colonies should be examined as set out previously and an assessment made of each. Many problems turn up at this time of year, and if you keep bees long enough all of them will turn up at some time or other. However beginners do not usually have too much trouble in their first few years.

Dead colonies

Dead colonies are obvious – no bees flying and only dead bodies in the hive when opened. Combs can be used again as long as they do not contain larvae, or the scales of larvae, which have died from foulbrood disease, see Chapter 7, page 76 et seq. However, I would not advise using them again if they contain large patches of dead brood, large numbers of dead adult bees who have died head first in the cells or large areas of mouldy pollen. It is not worth the work which bees have to do to reclaim combs of this type. It is better to melt them down and reclaim the wax and give the bees new foundation. In any case before the combs are used again they should be fumed with acetic acid as directed for Nosema control. This method of fumigation is effective against most of the diseases except AFB and should be routine practice upon all spare used combs before they are reused.

Live colonies

Examination of the brood in some colonies will show the high domed drone capping over many of the pupae in worker cells, the cells being widened at the mouth and lengthened, thus deforming the comb. This indicates either a drone laying queen or laying workers. Colonies with this condition may, or may not, be worth recovering; it rather depends upon the number of workers still alive.

Colonies may be small in spring because of being left with inadequate winter stores, poor or defective queen, or Nosema which is a microscopic gut parasite. If necessary a good feed of about two gallon of syrup should be given. Once the stores position has been dealt with any further reluctance to build-up in population size will be due to one of the other two problems mentioned above. How are these recognised and differentiated? Think of the factors that are causing the problems. In the case of the failing queen she is not increasing her rate of egg laying fast enough and therefore her brood area is small compared with other colonies which are doing well. Her worker bees are however living their full length of life. In fact their lives will tend to be extended a little because of the small amount of brood to feed. The picture one gets therefore is a small brood nest with a large number of worker bees looking after it. A colony suffering from Nosema will be opposite to this. The queen will be laying as many eggs as the population of workers can look after, whilst the diseased workers will lose up to about a third of their life span. The picture seen by the beekeeper will therefore be again a small brood nest but this time very sparsely covered by worker bees. The failing queen should be replaced and the diseased colony treated, see 'Queen Introduction' page 115 and 'Nosema' page 83. Average colonies can be built up by ensuring they are never short of stores and are supered ahead of requirements.

Summer

Apiary work in the summer, once the colonies are fully built up, will be the need to add supers and deal with colonies making queen cells ready for swarming. Supering is most important. The bees should never be allowed to get into the position where they are standing on all the comb in the colony, and certainly not into the position where there is insufficient room to store fresh nectar. Congestion of this sort will considerably increase the chances of swarming. Supering is dealt with on page 59 and Swarming on page 65 et seq. Routine inspections should be carried on until after the apiary has had its swarming spell. Experience shows that the colonies in an apiary all try to swarm, if so inclined, at about the same time and then it is over for the year. The odd one or two may swarm towards the end of the season but it is not worth routinely lifting a lot of honey to look for them. Some can be discovered by putting about four frames with wax foundation in the top super. If the bees are drawing this then there is no problem but if they have started to draw it and have stopped and left the cells half-finished they will probably be making queen cells in readiness to be off. If the foundation has not been touched then this can denote many things: the flow may have ceased, the colony have decided the year is over, the queen is beginning to fail or they may be thinking of swarming – you can take your pick and should go into the brood nest to find out.

Heather areas

Beekeepers who will have, or are going to move their colonies to a heather flow at the end of the season need to get ready for this earlier in the year. One cannot expect to get a second period of work out of colonies which have already worked one flow and settled down towards winter. Prospective colonies going to the heather should have a young queen raised in the current year. Make up a small three comb nucleus with a well nurtured queen cell from a swarming colony or purchase a queen. See 'Queen Replacements', page 116. The nucs should be built up during the season until the move to heather is to be made. The supers of the major colonies are cleared, taken off and extracted. These colonies are inspected, the queens found and killed and a nucleus is united to each. Once uniting is complete the combs in the brood chambers are rearranged putting sealed stores on the outside, the sealed brood in the centre and the eggs and young brood on both sides of the sealed brood. The advantage of this arrangement is that the eggs and young brood will fill these outside combs for about three weeks thus preventing the bees putting honey in them, while the brood in the centre combs will emerge and the queen will relay them. Without this arrangement the queen will begin to reduce her egg laying on the moor and will fail to relay the outside combs which will be filled with honey which the beekeeper will not be able to take away. The extra brood and bees from the nuc and the extra vigour of a young queen will provide a much more adequate and eager forager force to deal with the conditions on the heather moor. When the colonies are rearranged care should be taken that they have sufficient stores to last them a week or more. They are given at least one empty super and moved to the moor. The beekeeper should visit the colonies on the moor to ensure they are not starving or in need of more supers.

Harvest

At one time most beekeepers used to leave their honey on the hives until the end of July or the beginning of August and then take it off in one go. Those in heather areas had another flow to come later, but now if you are in an area with oilseed rape then it is necessary to take the rape honey off as soon as the field goes green and the flow is finished or the honey may crystallize in the cells of the comb. Once the honey flow has finished you can clear the workers from the supers and take the honey off to be extracted and stored. Clearing bees is dealt with on page 90, and Extracting on page 93 et seq. When you clear the bees, or put clearer boards on, you should examine the colonies and make sure they have sufficient stores in the brood chamber and that you are not taking away all their food. At the end of the season replace any poor queens and any that have done their two years. In this way colonies will be ready to be fed down for winter before the first frosts arrive. With feeding finished the year is over and indeed the foundations for the next have already been laid.

6. Swarm prevention and control

Swarming

All bees are genetically programmed to swarm as this is their only means of natural colony reproduction. Some bees are more 'swarmy' than others.

In beekeeping parlance the statement 'this colony is swarming' can have two different meanings. It may mean the colony is pouring out of the hive into the air to form a swarm, or it may mean that the colony is making queen cells and may be going to swarm out in the near future. When a colony is making queen cells and decides to swarm it will usually leave the hive when the first queen cell is sealed, or perhaps earlier. To hold them up for about another week the queen's wings are clipped so that she can't fly. With a clipped queen the swarm does not usually leave before the first young queen is emerging. If the bees attempt to swarm the queen cannot go with them. She either stays in the hive or if she tries to accompany the swarm she falls to the ground and is lost. In either case the bees return to the hive after hanging up for a short while. The colony will eventually swarm when a new virgin queen emerges from one of the queen cells. Thus clipping the queen gives the beekeeper the assurance that the bees are unlikely to swarm until sixteen days after queen cells are started, and the certainty that if they do, only the queen will be lost. It is easy to replace one queen but impossible to replace 30,000 workers lost in a swarm. Beginners will probably have to get a competent beekeeper to clip their queens until they have attained the confidence to pick up a queen and handle her, for technique see 'Clipping the Queen' page 109.

Swarm prevention

The three main causes of swarm preparations are a 'swarmy' strain of bees, aged queens and congestion of the worker bees in the hive. To reduce swarming therefore one has to eliminate these causes. If you have a 'swarmy' strain of bees then you will need to replace the queen with a better strain from another beekeeper. Age of queen is a vital factor which one can do something about by ensuring that the queens in your colonies are changed at the end of their second season. How they are changed will depend upon your particular circumstances. New young queens can be bought in or they can be produced, under careful supervision in your own colonies. The latter course is better as it involves one in queen rearing, one of the most interesting sides of beekeeping. The third, and very important, aspect of swarm prevention is preventing congestion by supering at the correct time. Supering was dealt with on page 59 when discussing building up the nucleus. The same applies to an established colony. They should never be allowed to get to the state where they are standing on and using every bit of comb in the hive. Keep ahead of them until the end of the season by putting on more supers. You will need at least two to three supers per colony in most years and in a good year, when everything goes right, this will not be enough. It is easier if you do not exceed four supers on at one time. Take one off and extract when the fifth goes on, otherwise it gets a bit heavy when you examine the colony. Giving plenty of entrance space also helps. Open out the block when the weather becomes nice in the spring and the colony has a super on and take the entrance block right out when the second super goes on.

Beekeepers are advised to inspect their colonies at least every 10 days. This is called by beekeepers 'the ten day inspection system' and is reliant on having a queen whose wings have been clipped. See page 54 et seq. This chapter documents the stages in inspecting the colony. For a summary of how the system works see page 119.

Examination during swarming season

Once colonies are built up, and one has moved from asking question 2a to 2b, queen cups will be built by the bees and these should be examined by tearing out their side with the corner of the hive tool. To start with the cups will be matt on the inside but they will soon be polished and eventually an egg will be laid in some of them. I would not worry about these eggs. Often they are eaten by the bees and the queen relays later. Once the queen cells contain larvae and the bees begin to feed them then we must assume they are going to try to swarm. In fact they may give up the whole process of queen production at any time but unfortunately we cannot say which colonies will give up and which continue so have to treat them all the same.

Finding the queen

Many of the manipulations start with the need to find the queen for some reason or other. This is a major problem to the beginner and less experienced beekeeper, but it is something that must be learnt. When the weather is fit to keep the colony open for a while, temperature above 60°F / °15°C, and reasonably wind still, the beginner should practise finding the queen.

What to look for

A bee which is double the size of the workers, which stands out above them on the comb, is more 'spidery' in appearance having longer legs. We are looking for the queen of a colony, a mated queen, in which case she will be in full lay and therefore heavy with eggs and moving slowly and sedately around the comb, unless you get her on the run when she will move around fast.

How to look

Open the colony quietly using as little smoke as possible consistent with retaining control of the colony. Look for her in a systematic way. Start with the outside frame. Take it out of the hive and look at the bees on it allowing your gaze to travel around the outside of the frame first and working in a spiral to the centre, turn the frame over to look at the other face of the comb. While turning over look at the bottom, top and side bars to make sure she is not walking around from one side of the comb to the other. Then examine the second side in the way you did the first one. Reverse the frame again and examine the first side once more, and then put the frame down. This being the first frame it is generally left outside the hive, leaning against the front of the hive, or put into a nucleus box for temporary storage. Subsequent ones will be put back into the brood chamber. Repeat the process on each frame, until you find the queen or get to the other side of the hive without finding her. On this first time through the frames do not take a long time looking at each frame. Let your eyes pass over fairly rapidly, because it is highly likely you will see the queen more easily while the bees are relatively undisturbed. When you take a fresh frame from the hive, look at the face of the next frame still in the brood box, as you may see the queen walking around amongst the workers. Then look at the side of the frame in your hand looking first at the dark side, the side that was not exposed to the light when removing the previous frame. If you have not found the queen on the first pass through the box then start back through the brood chamber starting at the beginning again. Examine the combs in the same way but on the second pass take longer and move the bees about on the comb in case the queen is hiding under a cluster of workers. You can get the bees to move by shifting them gently with your finger or by blowing gently on their backs. In this way small clusters of workers should be broken up while watching for the appearance of the queen amongst them. If you do not find her on the second pass through it will not be worth carrying on looking because by this time the workers may be running about a bit, it will be more difficult to see the queen and you will only succeed in thoroughly disturbing the colony. However, if it is imperative that you find the queen then the following method can be successful. Take another brood chamber, or some other box, and put into it five of the frames from the hive arranged in two pairs and one singleton. The six remaining frames in the hive should be arranged in pairs, and in both boxes each pair of frames should be spaced apart from the next pair so that the light can get onto the outside faces of the comb. Leave open to the sky for a few minutes during which time the queen will usually creep into the dark between the frames in one of the pairs. Pick up both frames of each pair together and open them like a book and, if she is there, the queen will be usually be easily seen on one of the two comb faces. This is easiest done with two beekeepers, each taking and inspecting a combs. You will see at times suggestions that you should sieve the bees through a queen excluder to find the queen but I would not advise this method. It is a very drastic process and really needs to be done by someone who is expert in bee handling as poor queens may be small enough to get through the exclude. Queens can usually be found by the experienced beekeeper.

Finding virgin queens

There will be times, particularly after a colony has swarmed, when you will need to find a virgin queen. The method of looking is the same as above for the second time through the brood chamber. The difference is in what you expect to see. The mated queen moves fairly slowly and the bees part in deference to her, unless you have stirred them up too much. A virgin queen is slimmer and shorter. She rushes about, often fanning her wings, climbing over or burrowing under lumps of bees, pushing the workers about and often leaving a line of disturbance in her trail. They are more difficult to find but fortunately it is not often necessary to try, usually only when you have previously made a mistake of some sort.

Swarm control

Once the colony is making queen cell then some method of swarm control must be applied if you are not to lose a swarm, perhaps amounting to well over half your bees, the old queen and usually most of your crop of honey for that season. The simplest, and very versatile technique to use is the Pagden, or the 'artificial swarm', method. It must be realised however that it is necessary to have more equipment than just the hive the bees inhabit. You must have extra brood chambers and frames to deal sensibly with swarming, no matter what form of management you use.

Artificial swarm technique

You need to have prepared an extra brood chamber filled with frames containing comb or foundation, and something to act as a floor, crown board and a roof – preferably the proper things. When you find a colony making queen cells you lift it off the stand to a position about two feet to one side. The new brood chamber is placed on the new floor in the exact position from which the colony has just been moved. It is filled with ten frames containing drawn comb or foundation, leaving a space in the centre to take the eleventh frame. If only foundation is available a queen excluder should be placed on top of the new floor to stop the bees with the queen absconding. This can be removed as soon as the bees have drawn some of the foundation and the queen has started to lay in the new frames. The colony is searched until the queen is found. The queen together with the frame of brood on which she is found, is placed into the centre of the new brood chamber. This frame of brood should be searched for queen cells which are squashed. Any supers are then placed on the new brood chamber, above a queen excluder, and the inner cover and roof is put on. The hive now on the old stand has the queen, one frame of brood and ten empty combs, the supers and the super bees. The latter will be augmented by the flying bees from the old brood chamber, who return to their usual position on the stand as soon as they fly out from the new position. The old brood chamber should be covered with a crown board, and have a roof put on. This unit and can be left alone for a week. It is a good idea if you have, and only if you have, some almost fully grown but unsealed queen cells in the old brood chamber, to kill any sealed queen cells thus ensuring no young queen will emerge for at least eight days. Seven days after the artificial swarm is made the colony on the old stand should be examined to see the queen is laying well and expanding her brood nest showing the colony has given up trying to swarm. The old brood chamber with the queen cells should be moved to the other side of the colony with the queen. This will mean that the thousands of bees which have started to fly during the preceding week will fly out and return to their site which will now be empty. The nearest hive being the colony with the queen, the bees will enter and be accepted by their erstwhile hive mates. Thus the queen's colony will again be augmented by another lot of young flying bees. The queens are left to emerge in the old brood chamber, which as it has had the population drastically reduced will normally not to try to swarm. Young queens emerging will automatically be reduced to one by the first virgin queen to emerge. However if you wish to be absolutely sure that no swarm will come out you could reduce the queen cells to one at the time of the second shift, see 'Reducing queen cells to one' page 72. The surviving virgin queen will mate, and start to lay in about 20 days after the artificial swarm is made. Once the new queen is laying the old queen can be killed and the old brood chamber with new queen united with the new box, now dequeened. This is done by moving the old box on top of the new as described below.

Uniting is best done by the 'paper method', see 'Uniting' page 120, using a queen excluder between the old and new brood chambers and another queen excluder between the old brood chamber and the supers to prevent the queen going up into the supers. A week later the combs of brood can be sorted into the bottom box accompanied by the new queen. If there should be more than eleven frames of brood the extras can be left in the brood chamber above the supers until the brood has all emerged. Then the second brood chamber can be taken away, the honey extracted, and stored until next year. This is the artificial swarm method of control used without making increase.

The same system can be used to increase the number of colonies by moving the old brood chamber away to a new position on the second shift and building it up into a new full sized colony before winter. The technique can also be used to produce a small number of new queens instead of just one, see 'Queen Replacements – production of' page 116. The artificial swarm method is extremely efficient providing you have the extra set of drawn comb.

If you have to use all foundation then there is a possibility the bees will be reluctant to draw the foundation if the weather is poor and forage scarce. The bees may continue to make queen cells on the single frame of brood. If the queen is still present the new queen cells should be killed and generally this will put things right. Should the queen be already gone with a swarm then take a frame of brood with a sealed queen cell from the old brood chamber and put it into the centre of the brood nest. This will save time as the queen cell will be nearer to emerging than any of those started after the colony was artificially swarmed.

If using only foundation is it is better to leave the supers with the original colony with the queen cells and feed the artificial swarm to enable the colony to draw the foundation. If the supers were left with the colony it will not be possible to feed as some sugar syrup may be stored in the supers. As soon as the brood comb is drawn the supers can go back onto the artificial swarm.

Lost clipped queen

A colony with a clipped queen may swarm out and lose her onto the ground after which the bees will return to their hive. Usually this only happens if you have missed an inspection or failed to see queen cells in the colony. You should detect that the queen is lost by the fact that there are no eggs. Remember eggs are always the first things you look for in the colony. Complete absence of eggs will only occur if the queen has been gone over three days, and the last eggs she laid have had chance to hatch, but usually the clipped queen lays very few eggs in the last few days before she goes and therefore if you have to really search to find eggs the chances are that she is gone. Another indicator of a lost queen is that the bees are more 'touchy' and on edge than usual when being manipulated. Fortunately using clipped queens it does not matter if you read the situation incorrectly, thinking she has gone when she is still there. This will do no harm as she will be gone very shortly. Once the queen is gone the colony will swarm again as soon as the first virgin queen is able to fly, usually in two or three days, so it is necessary to deal with the colony as soon as possible. The answer is to remove all the queen cells but one and leave this one to emerge, mate and start laying. See overleaf for details of method. It may take a long time for the new queen to mate from a large colony.

Colony with an unclipped queen

With any luck if you examine your colonies every seven to ten days you should find the ones with queen cells before they have lost a swarm and the queen is gone, in which case you can artificially swarm them as detailed above. However, sometimes they will beat you and swarm out. If someone has an eye on the apiary most of the time they may see the swarm flying and know where it has settled and formed a cluster. In this case you can take the swarm, see page 74, and then deal with it in a similar way as doing an artificial swarm. That is shift the brood chamber to one side, about 2 feet, onto a floor with a cover and roof. Place a floor and a new brood chamber on top on the stand where the swarm came from. Fill the brood chamber with frames of foundation and hive the swarm into it. The old brood chamber can be treated in exactly the same way as the old brood chamber of the artificial swarm. The only real difference in the two cases is that the natural swarm is given all foundation.

Reducing queen cells to one

The first thing to do before reducing the queen cells to one is to go through the colony and pick out a good queen cell, the one you are going to save. Look at the queen cells and pick one which is about 1 to 1¼ inches (3 cm) long, no longer or it may be a duff one, and about ½ inch (12mm) in diameter. Smooth queen cells are usually defective in some way. Before you finally decide to keep this cell just lightly scrape the bottom end of it with the hive tool. If the end comes off at the slightest touch the occupant has already emerged and the cell is useless. An occupied cell will still have the cocoon intact and nothing will remove its end. The chosen cell may be marked by placing a thumb tack into the top bar above it. The comb should then be carefully searched to ensure there are no other queen cells left on it. Any found should be squashed. There is only one way of being certain that you will find every queen cell and that is to shake the bees off the combs into the bottom of the brood chamber, being careful not to shake the frame with the queen cell which is being saved.

To shake a comb the frame is held, with the fingers curled under the lugs, with about a quarter of its depth still below the top of the brood chamber and then shaken up and down a couple of times without bashing it against the sides of the hive. This comes with practice. I generally shake down so that my fingers below the lug hit the top of the sides of the brood chamber. This provides a jerking stop which dislodges all but the youngest worker bees. The comb will now be only thinly covered with bees and queen cells can be seen easily. Open queen cells can be squashed; be certain you kill the larvae. Sealed queen cells are grasped by the fingers, without pressing on the cell very hard, and the bottom end torn off. If you remember the queen larva only spins a cocoon inside the lower half of the cell so it is above this that the cell breaks away. You will now have in your fingers the bottom end of the queen cell containing the living contents. Look at this and decide what stage it is at. If the contents is a white tail then it is a pupa and can be squashed, if it is a pale yellow bee abdomen then it is too young to be kept, but if on the other hand it is a dark yellow or brown abdomen then let it out onto something and put her into a cage. If the young queen can walk she is alright to save. If on the other hand the queen cell is empty or shows the head of a bee, a worker if you look closely, then a queen has already hatched from these cells, and the worker was clearing up the remains of the queen's food, when some of her sisters resealed the lid back on. The workers often reseal an empty queen cell. If you have shaken the bees and killed all the queen cells but one without finding a virgin queen old enough to be walking then you can finish the manipulation and leave the selected cell to emerge and mate over the next ten or more days. If, however, you have found virgin queens which are able to walk then let one go free in the colony and kill the selected cell which was going to be kept as this will not now be needed. The gain is a few days in most cases. There is a danger when leaving more than one virgin as they can fight and damage each other. Only leave one unless you have no choice. They will not swarm unless there is a back-up queen cell available.

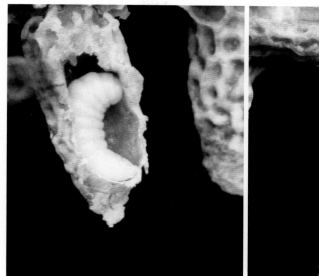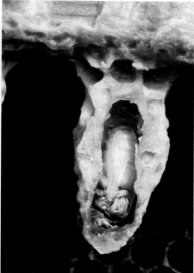

Queen larva and pupa shown in good queen cells.

Colony left with selected queen cell

The beekeeper will not be sure of the exact age of the queen cell left in the colony so the queen may emerge any time during the next eight days. Therefore if the colony is examined seven days later it is possible for the queen to be still in the cell and a beginner may worry that the queen cell is a dud although in fact it is quite alright. Usually however, she will have emerged and the cell will bear evidence of this by having an open end. At the time the queen cell is left the beekeeper will not necessarily know the age of the brood in the colony. If there are open larvae then it is possible to decide at what time the queen left and the age of the brood will be known. There are times, particularly when routine inspections have been missed and all the brood is sealed the exact age is difficult to determine. If at the next examination no brood is found in the colony, the first thought of many beekeepers is that the colony is queenless. This may or may not be true but in my experience it is very unusual for a colony to become queenless in the active season unless the beekeeper has made a mistake in some way and prevented the bees from keeping a queen or a queen cell. Most likely there is a virgin queen in the colony who has not yet started to lay. When a large colony is left with a queen cell it will often be some weeks before the new queen begins to lay, so there is little need to worry before about three weeks have elapsed after the queen cell was left. After two weeks it is time to make sure the colony has a queen and this can be done by putting in a 'test comb'.

Test comb

A test comb is any comb from another colony which contains eggs and very young larvae. It is essential to have young larvae on the test comb because if it contains only eggs these will often be eaten by the bees. This comb is put into the colony to be tested for queenlessness and if in three or four days time they have started to make queen cells then they are queenless. If they have not started queen cells then they have a queen. In the swarmed colony this will be one that has not yet started to lay. This method of testing can be used on any colony suspected of being queenless at any time during the active season. The only time this does not work is in a colony which has swarmed within the last week. In this case the colony will often make cells to continue the swarming process and swarm with the virgin that is already in the colony.

Collecting a swarm

Swarms are usually quite docile and easy to handle. When they come out they carry three day's supply of honey inside their honey crop and are happy. If, however, they have hung up in bad weather for a few days, and used up most of the honey, then they can be quite irascible. Always put on at least a veil before going to deal with a swarm and have the smoker lit. Swarms are usually collected into a straw skep or a wooden carrying box with a removable lid. When using a skep it is best to have a small sheet of cloth, a minimum of about five feet square, which can be laid on the ground and the skep full of bees placed on it so that when they are to be taken away the sheet can be folded over the skep and the corners tied, also a piece of string tied around the cloth on the skep to prevent the bees climbing up the outside and escaping. The advantage of the skep is that it is lighter in weight and somewhat flexible so that it can be pushed into awkward corners more easily than a wooden box. The rough surface of the skep will give the bees something to hang onto. In an emergency cardboard boxes can be used to collect swarms, but care must be taken that they are kept dry or they will collapse under the weight of the bees.

All swarms can be collected by one of the three following methods. If the swarm has clustered upon a thin branch of a tree, which is whippy enough to be easily shaken, then the skep, or box, is held close under the cluster of bees and the branch shaken vigorously to dislodge most of them into the skep. The skep is inverted onto the sheet with a small stone propping it up slightly at one side so that the bees can get in and out. If a box is used its lid is put on and it is stood on the ground with the entrance open, close to where the swarm was hanging up. Try to put them in a place out of the sun, or shade them with something to prevent them heating up. The branch they were on is now heavily smoked to make the bees fly and join those in the skep, many of which by this time will be outside the entrance fanning and scenting to call in any bees which have not yet rejoined the swarm. If the queen is missed when the swarm is shaken she will usually join the fanning swarm in the skep if the branch is well smoke. If she is missed altogether the swarm will leave the skep in about half an hour and hang up again, sometimes moving away for a short distance. The queen must be with the swarm for it to stay anywhere for very long.

Usually the swarm has not clustered in such a convenient way but is on a thick branch, or maybe in the thicket of a hedge, neither position allowing it to be shaken. A swarm has two main instincts, to move upwards and to go into the dark. If therefore it is possible to put the skep over the top of them they will move into it if you can start them off. Usually a little smoke will help but better is to scoop a few bees from the swarm and throw them into the inside of the skep. Those that pitch will usually start to fan and scent and this will call in the main body.

Finally the swarm may not be in a position where they can be shaken nor can a skep be placed above them. In this case it is usually possible to brush them into a skep or to brush them onto the ground and put the skep over the top of them afterwards. One or a combination of these three methods will allow most swarms to be taken.

Warnings

Beware using an unfamiliar ladder as it may not be as sound as it looks. If you are up a ladder taking a swarm ask, if necessary, another beekeeper to hold the foot of the ladder. Bees landing on a non-beekeeper will often cause them to leave their post very quickly, with disastrous results. Remember, if you are leaning out from the ladder to shake a swarm that it will weigh several pounds. When it drops into the skep this must be allowed for or you may drop the skep or fall off yourself. The weight is a surprise to most people as they do not think of insects as having noticeable weight. Swarms that have entered a hole in a wall or gone down a chimney have ceased to be swarms and you will have great difficulty in collecting them. In a wall the bees could only be got out quickly by opening up the cavity they had gone into and this would cause structural damage which you would be expected to repair. My advice is to keep away from jobs of this kind. Remember however that it is imperative for the continuance of good relations with non-beekeepers for all beekeepers to do their utmost to prevent loss of swarms from their own hives. All beekeepers should be willing to take any stray swarms they come across, or are asked to deal with, to prevent them being a nuisance.

7. The Honey Bee's Diseases and Pests

General

The honey bee like all living things will have various organisms attacking it and causing disease and at times its bodily functions will go astray and cause trouble. From the practical beekeeping point of view we are not worried by the problems of individual bees. It is only when the disease affects a large number of bees that we notice this and start to deal with the problem. Most of the diseases of the honey bee will not kill the colony but they will reduce its value as a collector of honey. Many are endemic in the colony and only flare up into importance when the colony's resistance is lowered by some adverse condition, often brought about by the beekeeper's manipulation. Honey bee disease divide very easily into those which attack the brood and those which attack the adult bee and will be dealt with under these main headings.

Notifiable diseases

There are legally enforceable regulations aimed at protecting the honey bee population of the UK from disease. Two of these are brood disease, the foulbroods AFB (American Foulbrood) and EFB (European Foulbrood). The same legislation covers the infestation of colonies with two exotic pests, a mite, *Tropilaelaps* and a beetle known as the Small Hive Beetle *Aethina Tumida*, neither of which has reached the British Isles at the time of writing. The mite *Varroa destructor*, previously notifiable is now so widespread that it was removed from the legislation in 2006. The Food and Environment Research Agency (FERA) administer the legislation through bees officers at the National Bee Unit (NBU), Central Science Laboratory, Sand Hutton, York YO41 1LZ. The NBU produce a number of useful booklets free of charge and it has a useful web site accessible through the internet site: https://fera.defra.gov.uk/beebase. Beekeepers are advised to register with BeeBase which gives much useful information.

Two notifiable diseases not yet found in the UK at the time of writing (2013):

Tropilaelaps species

The natural host of this parasitic mite is the giant Asian honey bee *Apis dorsata*, but the mite will readily infest colonies of other honey bees in Asia. Its life cycle is similar to that of Varroa in that it affects both the adults and the brood, but only breeds on brood in brood cells. The *Tropilaelaps* mite is the same colour as varroa, brown, but smaller and elongated.

Small Hive Beetle – *Aethina tumida*

This beetle is dark brown, about the size of the small finger fingernail. It enters beehives and lays clusters of eggs, which hatch into larvae similar to wax moth larvae, but with a row of spikes running along their back. They destroy honeycomb and leave the honey slimy and smelling of rotting oranges. The NBU should be notified immediately if you see any sign of a small beetle in a hive.

FERA produces free booklets which will give you more detailed information on both these infestations.

Notifiable brood diseases

American Foulbrood – AFB

This disease is caused by infection with a spore forming bacteria called *Paenibacillus larvae* subsp. *larvae*, the spores of which are extremely small but very persistent, remaining infectious for over thirty years and probably for much longer. The spores are brought into an uninfected colony when foragers rob another colony which has the disease. The spores do not appear to affect the adult bee. Spores are fed to the house bees in the normal passing of food from one bee to another and are eventually fed to the larvae by house bees doing nurse duty. The infected larva dies just after the cell is sealed and rots down to a sticky, coffee coloured thick slime. During this process the cappings of the sealed cell becomes dark and moist looking, sunken and often perforated. Worker bees recognise a problem and partially remove the capping. If a matchstick is pushed into a cell with a sunken moist capping it will pick up some of the remains of the larva, which will pull out as a sticky rope. This 'ropy' stage is diagnostic of AFB. The rotting larva eventually dries to a hard black scale lying horizontally along the lower walls that form the lower 'V' of the cell. It is firmly attached to the wall and cannot easily be removed by either the bees or the beekeeper. The firmly attached scale is diagnostic of AFB and the scale will remain on view for as long as the comb lasts. During the ropy stage and after, the bees will try to clean out the cell and will get infectious spores on their mouth parts, which will eventually infect other larvae. As these larvae die and form scales still more cells will be taken out of use and will remain empty as the queen will not lay in cells with scales. Thus a comb of sealed brood with lots of empty cells amongst the capped brood could be showing signs of AFB. This is the 'pepper pot' stage. A comb showing 'pepper pot' appearance should always alert the beekeeper to look at it very carefully for the presence of AFB scales. The disease can be confirmed by finding scales in the open cells. These can be seen by looking into the cells from over the top bar with the light coming from behind. Further confirmation can be obtained by using a kit similar to a pregnancy test – a specific lateral flow device. The presence of the disease should be reported to the local Regional Bees Officer or directly to the NBU, at Sand Hutton. A Bees Officer will come and confirm the disease and destroy the colony, burn the infected comb and flame the equipment to kill the spores present on everything which has contacted the colony. In Britain, as the law stands colonies with this disease must be destroyed. In other countries a colony with AFB can be treated with antibiotics, but in my opinion, treatment is uneconomic in both time and money. It is interesting that in countries using treatments the incidence of the disease is far higher than it is in the UK. Do not worry about this disease. You can keep bees on a large scale for thirty years and never have a case. Bees Officers have been inspecting colonies for the disease and the incidence is very low. The incidence could be reduced even further if all beekeepers kept their eyes open for the disease and reported it immediately.

European Foulbrood – EFB

The names of the two foul broods have no significance at all, the geographical part means nothing as they are both distributed worldwide and they are quite dissimilar diseases, so the fact that they are both called by the same name, foulbrood, is equally meaningless. EFB is caused by a non-spore forming bacterium called *Melissococcus plutonius*. There is much we do not understand about this disease, particularly its epidemiology, what its incidence is in colonies and how it is spread from colony to colony being two major points of ignorance. We do know that when the bacterium is present in the colony, and in the larvae, it can be totally invisible during normal routine methods of inspection. The bacterium is fed to the larvae by nurse bees with contaminated mouth parts. The bacterium multiplies in the stomach of the larva living on the gut contents. Providing the larvae receive sufficient food for their requirements, as well as sufficient to supply the needs of the bacteria, then the larvae will complete their development, although the resulting adult bees may be considerably lowered in vitality. The surviving infected larvae void the contents of the gut in the cell in the normal way, thus releasing thousands of bacteria to be cleaned up by the nurse bees and spread to other larvae when they are fed. The whole process is invisible to the beekeeper as there is nothing to see. Only when the larvae are short of food do some of them die and the effects of the disease become visible. The larvae die in the curled up stage before the cell is capped, often turning yellowish in colour and usually moving into quite unnatural positions in the cell. After death the larvae appear 'melted down', as though they were made of wax loosing the normal segmentation (Michelin man appearance). Finally the remains dry out to form a dark brown scale which can lie in any position in the cell. During the whole period the larva or the scale can be easily removed from the cell by either the house bees or the beekeeper. In cases where the infected larvae are fed sufficiently to survive to pupation, but not enough to complete their development, they die after the cell is capped. The signs of disease become similar in appearance to that of AFB with perforated, sunken, moist looking cappings. In this case however the cell contents will not 'rope' in the same way that AFB does. This is a notifiable disease and you should contact a Bees Officers locally or the NBU. The disease can be confirmed on the spot by using a specific lateral flow device. In the past colonies were treated with an antibiotic. The trend now is to do a 'shook swarm', where the bees are shaken onto fresh frames containing new foundation. This is detailed in Chapter 10 on techniques. This reduces the danger of any re-infection from the old comb or antibiotic getting into the honey. With severely infected colonies, and a positive result using the lateral flow device, the colony can be destroyed by burning. The Bees Officer has the power of destruction, but the beekeeper may insist on having a comb sent to the NBU for laboratory confirmation. At present it remains illegal for beekeepers themselves to treat the colonies with antibiotics in the UK.

In the case of suspected AFB or EFB do not move any bees or equipment in or out of the apiary until the Bees Officer has been and either cleared or confirmed the disease. The Bees Officer will advise you on how to proceed.

Other brood diseases

Sacbrood

This disease is caused by a virus, which prevents the infected larvae from casting its final larval skin and turning into pupae. This happens after the cell is sealed, but bees usually remove the capping and the dead larva is seen as a dark 'toe' pointing upwards in the mouth of the cell. This is called the 'Chinese slipper' stage, and once seen it is easily recognised. Larvae dead of Sacbrood are easily removed from their cells. Sacbrood is very widely distributed, but is usually not a great worry, only a few larvae in a colony being affected. Some colonies seem to suffer considerably and these should be re-queened, preferably from a different strain of honey bee as there is possibly an inherited susceptibility to the virus.

Chalkbrood

Chalkbrood is caused by a fungus called *Ascosphaera apis*. The fungus produces many thousands of sticky spores, which are probably present in all colonies of honey bees and these are fed to the larvae by the nurse bees. Normally healthy larvae do not succumb to the disease. Only in larvae which are under stress for some reason, the spores germinate and the larvae die. Reasons for succumbing to the fungus are slight chilling of the larvae, probably malnutrition and high carbon dioxide content in the air of the cell. Once the spores germinate the fungus grows through the body of the larva and envelops it, except for its head. In the early stages it fills the cell and looks like cotton wool with a yellow spot on top, but then dries out and shrinks down to a hard mummy, which is loose in the cell and can be removed easily. This mummy can be white in colour, or turn very dark blue as the fruiting bodies of the fungus are formed. These will contain the spores which will carry the fungus forward to its next generation. Chalkbrood is not a great worry, it flares up in some years, but on the whole it is only a few larvae that suffer. There is no treatment, bad cases should be re-queened and the beekeeper should avoid manipulations which are likely to chill the colony, particularly in the spring.

Stonebrood

This is a fungal disease caused by *Aspergillus* species. It is widespread over the world, but few colonies suffer from it. There have only been two cases in the British Isles as far as is known. The larvae die and assume the appearance of yellow sandstone. There is no treatment. Combs containing Stonebrood should not be sniffed as the fungus can cause infection of the windpipe.

Varroa

The disease Varroasis is caused by the mite *Varroa destructor* which was originally a parasite of the Asian honey bee (*Apis cerana*). Despite import restrictions it was found in the UK in 1992 and had obviously been here for some time. This mite differs from other bee diseases in that it affects both adults and the brood, but reproduces only in brood. Female mites hide on the underside of the adult bees, biting through the softer part of the bee's abdomen to feed on the bee's blood. The mother mite will enter an open brood cell just before the cell is sealed. She will then lay eggs, one male and a number of females, which hatch and feed on the blood of the developing pupa. Mating occurs in the cell, the male dying in the cell. Mother and daughter mites emerge with the emerging bee and enter another brood cell to continue the cycle. Mother mites prefer to enter drone cells as the longer drone development time allows more daughter mites to complete their development and emerge. The damage to the developing bee is not entirely related to the level of infestation, though if more than one mother mite enters the cell the developing bees usually die or emerge deformed. Virus infections are associated with varroa, and these appear to cause more damage to the colony than the mite itself. If left unmanaged or untreated the number of mites will multiply over a period of four to five years to the extent that the colony will not usually survive.

Mechanical and biotechnical methods of mite control

Beginner beekeepers should seek help with monitoring and treating colonies for varroa. There are various mechanical methods of removing varroa mites from a colony. I would recommend that all colonies should have an open mesh floor so that any 'clumsy' mite can falls through the mesh. It is important to monitor the number of mites falling through the mesh monthly during the active beekeeping season and during periods of treatment. The natural mite drop is an indication of when treatment is essential, see Beebase for further details. Another mechanical method is dusting the bees with about a tablespoon of icing sugar using a sieve or household flour duster. Some mites lose their grip and fall off through the mesh floor. This method needs to be repeated frequently to be effective. Drone brood removal can reduce the number of mites to some extent. Place a super frame, with worker cells, in the brood chamber at the edge of the brood nest in the spring. The bees will build a comb extension on the bottom bars to fill the gap, usually of drone cells. Once the queen has laid in these cells and the cells are sealed, the extension can be cut off and then melted down, destroying any mites in the cells. This needs to be repeated every nine to ten days, two to four times in the early part of the season.

The queen trapping method is very labour-intensive and is documented in the FERA leaflets.

Making up an artificial swarm is as detailed in Chapter 6. Once the brood has emerged in the parent colony, one or two bait frames containing unsealed brood are transferred from the artificial swarm to the parent colony. The mites, now all on the adult bees, will be attracted to the open brood cells. The combs are removed and destroyed once the cells are sealed, removing mites present in the cells from the colony.

Chemical treatments should now be used with caution. In the past the licensed treatment involved using synthetic pyrethroids impregnated into strips, which were hung in the brood chamber. Unfortunately the mites have developed a resistance to this medicament and it is no longer effective or recommended. The other licensed medicament is based on thymol, an essential oil occurring in thyme. The authorised method of application is 'Apiguard', a gel with added thymol, presented in a small foil tray. The tray is placed on the top bars of the brood chamber, gel upward, and surrounded by an eke or wooden frame to allow the bees access to the gel. The fumes will kill the mites present on the adult bees, but do not penetrate the capping of sealed brood. The tray is left in position for two weeks and a second tray applied for a further two weeks. In this way most of the mites will be removed. The treatment is given in the early autumn when temperatures are around 15°C/ 59°F. Other formulations of thymol are now available.

In addition beekeepers use oxalic acid in the winter when there is least brood in the colony and most of the mites are on adult bees. This is sold already mixed for a trickle treatment as a 'hive cleanser'. Another natural chemical is formic acid which should be used with great caution as it is highly dangerous to humans and may have adverse effects on the colony. It is now being packaged in 'beekeeper safe' containers and is available for general use with the assistance of experienced beekeepers. Formic acid penetrates the cappings and is thus able to kill the mite in the brood cell as well as mites on adult bees. Lactic acid is used by some beekeepers, but is very labour-intensive. Caution should be exercised before using any of these acids.

Treatments for varroasis are continuously evolving so advice should be sought from the NBU and local experts.

Viruses affecting brood associated with varroa

The honey bee suffers from a number of viruses affecting both brood and adult bees. It is now thought that some of these viruses have always been present in the bee population, but without causing any visible harm. The mechanism of activating these viruses or the transfer from bee to bee is still little understood. The most commonly seen is Deformed Wing Virus - DWV.

This virus is associated with varroa infestation. Bees affected by this virus emerge from cells with deformed or malformed wings and are removed from the colony by house bees. A severe infection will lead to rapid colony collapse with the queen and remaining workers trying to cover patches of dead or dying brood. It is essential to keep the level of mite infestation as low as possible. A colony with a high level of bees with deformed wings can only be saved by removing the remaining adults from the source of infection, the mites in the brood nest. This entails doing a 'Shook swarm' - see Chapter 10, page 117.

Parasitic mite syndrome – PMS

PMS is a term given to a number of signs of disease in a collapsing varroa infested colony. This condition can be confused with EFB as larvae die at the open brood stage of starvation. Large slabs of open and sealed brood, both healthy and dead, are seen with very few adult bees to tend to them: the newly emerging bees are often deformed and unable to take on nursing duties. It is accompanied by a heavy infestation of the varroa mite and virus infections. The beginner beekeeper should ask for help if they see any signs of dead brood or deformed adult bees.

Non-infectious brood conditions

Three non-infectious conditions, which cause the death of larvae, can be thought to be disease by the beginner beekeeper. They are important as they are often a direct effect of inexperienced colony manipulation. One should always be careful, however, that after diagnosing one of these conditions one does not cease to look and therefore fail to find a disease that is also present, possibly being masked by the condition.

Neglected drone brood

This condition is particularly found in conjunction with drone laying queens and laying workers where the brood nest is small and a high proportion of the brood are drone larvae. The bees seem to give up tending these larvae properly and they die, giving the appearance of an infectious disease. The dead larvae are yellowish in colour and watery in consistency. The remains do not rope and can be removed from the cell.

Chilled brood

Chilled brood is characterised by whole blocks of dead brood of all ages. The brood of the honey bee nest needs to be kept at about 95°F/35°C and if the temperature falls much below this the brood will die. A reduction of temperature can be caused by a sudden loss of bees when the colony has a large area of brood. The surviving bees will move up from the bottom edges of the comb and this will expose a crescent of brood along the bottom of the comb to the cold and death. The whole block of brood will die – that is the diagnostic factor. The loss of adult bees can be caused by disease, poisoning or moving too short a distance to a new site. The inexperienced beekeeper often induce chilling when making nuclei as old bees go home leaving too few to maintain the necessary temperature in the nuc.

Starved brood

This condition looks very similar to chilled brood except that the whole brood nest is suffering or dead, while some adult bees may still be alive but moribund. Usually the brood dies first and a lot of the larvae and pupae are sucked dry by the workers before they themselves succumb. Often the first sign of starvation of a colony are these husks of larvae being thrown out of the hive. The remedy is to pour a cup full of sugar syrup over, and between, the top bars of the frames for immediate resuscitation, and then put on a feeder with a couple of gallons of syrup. Starvation is something that should never occur during the active season. Regular routine inspections should be carried out to ensure that the bees have sufficient stores at all times.

Adult honey bee diseases

Nosema

Nosema apis is a microsporidian, a spore forming gut parasite, recently reclassified as a parasitic fungus. The spores are swallowed by the adult bee and germinate in the gut. The spore cell contents are injected into the epithelial cells of the gut wall and multiply there. Once the epithelial cell contents are consumed thousands of spores are formed and released into the centre of the gut to pass out with the bee's faeces. The damage to the gut wall's structure and function results in considerable reduction in the length of life of the bee. The effects on the colony will depend upon the percentage of bees infected. There is no visible sign of the infection in the individual bee. Usually the disease shows itself by an inability of the colony to build up in size in the spring. The disease rarely kills a colony, but the slow build-up makes the colony useless as a honey collector. In the late autumn and winter when the bees are unable to get outdoors they defecate on the comb, leaving Nosema spores on the surface of the comb. In the spring the bees clean up the comb ready for the queen to lay in, become infected and their lives are shortened. The colony fails to build up rapidly and the appearance of a brood area very sparsely covered with bees is apparent. The colony builds up very, very slowly until really good weather arrives and flying occurs almost every day. This allows the voiding of spores in the field, the latent infection in the hive is reduced and the colony builds up too late to provide a honey crop to the beekeeper. There is often sufficient infection left in the bees for spores to be present on the comb in autumn and the cycle repeats. This will occur particularly in years when weather is bad and bees are confined to the hive on many days, during which time those bees suffering from Nosema will defecate on the comb. This voiding on the comb is usually too small to be visible and has nothing to do with the condition known to the beekeeper as 'dysentery', although the latter will help the spread of nosema in the colony if it is already present. To help mitigate the effects of nosema, all combs that come out of a brood chamber should be fumigated with 80% ethanoic acid (acetic acid) before using again. Previously an antibiotic called 'Fumidil B' was given in syrup. 'Fumidil B' is an antibiotic used only for the treatment of Nosema in bees. This is no longer available to beekeepers at the time of writing.

Beekeepers should note any colonies failing to build up in the spring. Submitting samples of bees from a failing colony to a competent microscopist will discover spores of Nosema if they are present. Autumn sampling is in my experience of little value.

At present the only treatment available to beekeepers is to move colonies onto clean comb.

Nosema ceranae is a species of Nosema which originally infected *Apis cerana* – the Asian honey bee. This has crossed species and now infects our western honey bee. *Nosema ceranae* seems to be more virulent and colonies infected do not build up. These colonies may continue to decline during the summer months and are known to die with piles of dead bees outside the hive, having died over a period of time. The diagnosis is by microscopic examination. The treatment is the same as that for *Nosema apis*.

Acarine

This disease, or rather infestation, is caused by a mite called *Acarapis woodi* which enters the main breathing tubes, the trachea, of the bees. By piercing the tracheal wall it sucks in the bees' blood which is bathing the tubes. This reduces the length of life of an infected bee and there is some evidence that viruses may enter the wounds made by the mite so increasing the bee's problems. The effect of the disease on the colony will depend upon the percentage of bees infested and the presence of viruses. This can range from no visible effect to a carpet of hundreds of infected bees in front of the hive, which have crawled out to die, this latter being probably caused by a virus. The disease tends to be more prevalent where, and when, poor or wet weather confines the bees to the hive and the mite can migrate more readily from one young bee to another. There is no longer a legal treatment available in the UK, although some medicaments used to treat Varroa will also kill acarine mites.

Viruses affecting adult bees

Chronic paralysis virus (CBPV)

CBPV has infected bees before the advent of varroa. It shows up in several different ways, the most obvious being the appearance of a considerable number of dead bees in front of the entrance of the hive. This can be a large heap of dead bees, which makes beekeepers think that their bees have been killed by spray poisoning. There is however a difference. In the case of spray deaths this happens quickly, and usually it is not spread over several days; also every bee is dead or shows the other signs of poisoning see page 86. When paralysis strikes it does so over many days, so that the dead bees in the bottom of the heap will obviously have been dead longer than those on top and many of the latter will be moribund but still capable of some reaction if prodded. Other signs are hairless bees trying to get into their own hive and rejected by guard bees, before dying. Secondly obviously sick bees unable to fly, crawling up grass stems unable to fly, before dying.

Acute paralysis virus (ABPV) associated with a varroa infestation. Colony deaths infected by ABPV have only occurred since European honey bees have been infested by varroa. This virus causes sudden colony deaths and is also associated with winter losses. There is no treatment other than keep the level of varroa as low as possible.

Other viruses

There are a number of other viruses that affect honey bee colonies. Research is ongoing as to the damage these viruses cause.

Non-infectious conditions

Dysentery

This is a very ill understood dysfunction of the honey bee. It is said to be due to too much water in the bowel of the bee, which only provokes the question 'why is there too much water?' and we do not know. The condition is shown in the spring by daubs of brown excreta around the entrance of the hive, and sometimes similar stains inside the hive and on comb and top bars. There is nothing much one can do except give a feed of warm heavy syrup which at times seems to help. In series of years when bad seasons follows bad season, the incidence of dysentery can rise. During the winter when heavy dysentery occurs in the cluster it is usually fatal and the beekeeper is often unaware that anything is wrong until the colony fails to fly in the spring. As mentioned before this dysfunction is not caused by Nosema as in at least 50% of the colonies no Nosema spores appeared in the dysentery. Of course if the colony was also suffering from Nosema there would be spores in the dysentery and, if the colony did not succumb during the winter, more house bees would become infected as they cleaned the mess up in the spring. This condition may be caused by the feed given by the beekeeper. Syrup should be made up of white granulated sugar only. Baker's candy which beekeepers sometimes use for additional winter feed can be detrimental. It is safer to use the commercially available candy made specifically for bee feed.

Poisoning – natural

A number of plants have substances in their nectar which is poisonous to bees. The severity of their effect varies in different climatic and geological conditions and therefore do not occur at all in many neighbourhoods. The chief offenders are the Rhododendrons, Kalmia, some limes chiefly *Tilia petiolaris* and *T. orbicularis* and sometimes red horse chestnuts, *Aesculus carnea*, although the latter appears to have a greater affect upon bumble bees.

Poisoning – agricultural or pest control

The ill-advised or careless application of pesticides can be devastating to colonies in the area. Usually colonies which have been affected by spray lose their entire force of flying bees, and sometimes their queen, which will prevent the colony showing a surplus of honey for that season. Farmers have become far more conscious of the need to preserve the countryside and the animals in it, including bees, than they used to be. Also there are sprays that can be used which will do little or no damage to insects other than the particular pests they are designed to kill. Therefore the best way of avoiding spraying problems is to get to know your local farmers and discuss your problem with them. In most cases this will be quite successful in avoiding trouble. You should be informed of intention to spray, which may help you to take action to avoid damage to colonies. You may be able to shift them to another place for a while or you can help to protect them by covering them with straw or long grass, turning them into a haycock while the spraying is taking place. On no account must you shut colonies in by closing the entrance with the block or mesh. If it is hot weather they will die, and strong colonies will be killed even in cool weather. Poisoning can also occur if a pest controller has killed a feral colony in a chimney or wall cavity, failed to remove traces of honey and wax, and not sealed the entrance adequately. Bees will plunder the honey or a swarm may enter the abandoned nest and be killed. Finally if the colonies are affected and you lose your crop then you can claim damages from the farmer or spray contractor. Whether or not you decide to claim compensation, a sample of the dead bees and a report of the incidence should be reported to the NBU who will carry out a formal investigation and inform you what to do.

Pests of the honey bee

Mice – *Mus musculus* and *Apodemus sylvanticus*

Mice will enter hives in the autumn, probably attracted by the warmth. They will make a large hole in the comb and bring in leaves to make a nest. The bees do not attack them once they are in tight cluster, but seem to dislike their smell as they move the cluster away from the area the mice are and will not use comb which the mice have been on. The answer is to reduce the hive entrances to not more than $\frac{5}{16}$ inch (7.6mm) high slot when using an entrance block, or $\frac{3}{8}$ inch (9.5mm) holes if a sheet or metal with round holes is used to cover the entrance. This should be in place in late summer before the first frosts occur.

Woodpeckers – *Picus viridis*

The green woodpecker will learn that there is a meal in a beehive and will break its way in doing considerable damage. Fortunately not all woodpeckers know this trick so a beekeeper's colonies may not be affected for years, but there is always the risk if your hives are in a woodpecker area. Once they have learned to enter the hives the only way to stop them is to cover the hives with chicken wire or cord netting before the frosty nights arrive.

Birds

Many birds will take a toll of the bees flying from the hive, but not in my opinion of any great importance unless you live in an area where some birds, such as the bee eaters, specialise in this behaviour. The small numbers taken by swifts, swallows, tits and sparrows usually has little effect upon the colonies.

Wax moths – *Achroia grisella* and *Galleria mellonella*

Wax moths are pests of stored comb, or any comb not covered by bees. The caterpillars of the moths eat the comb usually working within a network of silk and will destroy a box of comb in a few days. They prefer comb which has been bred in and will therefore attack empty brood frames more readily than super comb. The answer is to store small amounts in an old freezer, or store it 'wet', see Storage of comb page 118. The greater wax moth, *Galleria mellonella*, will also enter hives occupied by bees. Usually the bees will not tolerate its presence and hunt it out or kill it. Some strains of bees will live with the moth and put up with quite large numbers in the hive. Strains of this sort should be got rid of as they are usually extremely unproductive. The single caterpillars of the moth will occasionally be found eating their way through the cappings of the brood. Their presence will be detected and a white line of pupae, which have had their cappings removed so that you can see the pupal heads. The white line is the silken cover with which the caterpillar replaces the capping it has eaten. Usually if this white line is searched along with the comer of the hive tool the caterpillar can be driven out onto the face of the comb and killed.

Braula – *Braula coeca*

Braula, or bee louse, is a small wingless dipteron, a member of the fly family, which lives in the colony with the bees. This insect has become quite rare since the widespread use of medicaments to treat varroa. It is about the size of a pin's head, red in colour and usually found riding around on the back of a worker. They do no damage as they are not bloodsuckers, but move down onto the mouthparts and join in when the bees are feeding each other. They are greatly attracted to the queen who will sometimes have quite large numbers, twenty or so, on her head and thorax. Braula lays its eggs in the cappings of the honey and the tiny grubs bore their way through the cappings leaving a trail like a 'leaf miner'. This does no harm and is only a worry to producers of sections and cut comb.

Wasps – *Vespula vulgaris* and *Vespula germanica, Dolchliovespula sylvestris*

Wasps become a nuisance usually starting in August at the end of the season when the wasp nest is breaking up, the mother wasp has died, and their new queens and males having emerged. If a colony is small wasps may rob it out. Any spit syrup or exposed honey will start robbing.

Hornets – *Vespa crabo*

The European hornet is rarely a problem to bees.

The Asian hornet *Vespa velutina* is not currently present in the UK, but was probably introduced to France accidentally in soil, flowers or fruit. It is a highly aggressive predator of other insects so does pose a problem to honeybees. See Bee Base for more information.

8. Harvest

General

When the beekeeper has to remove honey from the colonies there are a number of decisions to be made and a number of techniques to be used. In the first place it is necessary to decide when to take the honey off and then to decide if the honey is in the correct state 'ripe', the water content low enough for it to be storable and saleable. First the supers of honey have to be separated from the bees, the supers must be 'cleared', and there are several methods of doing this. Once removed from the colonies the frames in the supers have to have the honey extracted. The extracted honey once out of the combs must be stored properly and finally it must be bottled for use or sale.

When to take honey

Timing of harvest will depend upon the local crops and the wild bee forage plants which are within the area covered by the beekeeper's bees. If the early season is warm and the weather is good in an area where oilseed rape is grown the bees may forage up to two miles from the hive. It will be necessary to take the honey off as soon as the rape plants have mainly finish blooming. The beekeeper is forced into this because rape honey will crystallise in the comb if left any longer. On the other hand in an area where the honey produced will stay liquid for months, this is most honeys except those from cruciferous plants and raspberry, then it can be left on until the end of the season, unless you are short of supers. Then you will have to extract some and return them to the colony. It may be necessary to take the existing honey off when a new flower starts to secrete and you do not want the two honeys mixed. This could happen if you are in an area where the ling heather, Calluna vulgaris, comes into flower at the end of the season, or if you are going to move your colonies to the heather moors. In my experience most beekeepers wish to extract honey as few times as possible during a season.

Is the honey fit to extract?

Ripe honey should have a water content of less than 20%, preferably as little as 18%. At this water concentration it will not ferment or spoil during storage and can be legally sold. Most sealed honey will be at this level so sealed honey can be extracted at any time. It is the unsealed honey which is the problem. If there is still a nectar flow in progress, it will have a higher water content making the honey likely to ferment if extracted. Once the nectar flow has ceased the honey will usually be brought up to the correct concentration within a day but the cells will not be capped unless they are full. The usual test to see if the honey is ripe is to take a comb with some unsealed honey. Hold the comb over the top of the super by the side bars, with the comb parallel to the ground, give it a quick shake. If no honey, or only one or two drops fall out then the honey is fit to extract. If on the other hand a number of drops of honey splash out then the unsealed honey should be left on for a few more days. Possibly another flow may start and these combs will remain until the next extraction. However, if the nectar has come from a plant which produces a rapidly crystallising honey, such as oilseed rape, then the combs which are left behind to finish ripening should be removed as soon as possible or the sealed honey may start to crystallise in the comb.

Clearing the supers

Once you have decided to take the honey the first job is to separate the bees from the supers, or 'clear' them. This can be done in several ways but beginners are advised to use 'clearer boards' as this is the method least likely to cause trouble. Another method is 'shake and brush', which can be quite exciting if you are a bit tentative in your approach to the bees.

Clearer boards

Clearer boards are the same size as crown boards, and in fact a lot of crown boards can be used as clearer boards as well, the difference being that there are escapes built into or placed into the board so that the bees can pass downwards through it and not return. The escapes can be small metal or plastic appliances such as the well known 'Porter Bee Escape' which are fitted into a slot in the board, or the escape can be constructed in the board itself as in the case of the 'Canadian Clearer Board' or one of its modifications. The method used to clear a colony is as follows. First assemble all the things you will require. If the existing crown board is to be used as a clearer board then you will need something to cover the top of the supers in such a way that there is no chance of bees getting in from outside. The metal or plastic escapes should be examined to see that all moving parts are free and not propolized down. In the case of Porter Escapes the springs should be adjusted so that there is about $\frac{1}{16}$ inch (1.5mm) between the points and slotted into the oval slots in the crown boards. The channels in Canadian escapes should be clear and not obstructed by anything. The colony is smoked and the crown board is removed and the unsealed honey is tested to see if the top super can be removed at the same time as any fully capped supers beneath it. If the top super contains unripe honey, and is not going to be taken off at the moment, it is lifted off onto an upturned roof. The full supers are lifted off and placed on another upturned roof – or onto something else suitable if you only have one hive. The unsealed super is returned to the colony above the queen excluder as normal. The clearer board is placed on the unsealed super. The supers for clearing are placed on top, covered by another cover board and the roof. If all the supers are being removed at once, the clearer board is placed directly onto the top of the brood chamber.

The underside of the clearer board showing the mesh covering the escape hole, and the ¼ inch gap through which bees escape to the lower box.

The clearer board must be put on the hive correctly with the escape hole uppermost so the bees have to pass down through the gap made by the springs of the Porter escape or the small holes in the Canadian type.

If the clearing is done mid season say for oilseed rape it is wise to put an empty super below the clearer board to give the bees space. In this case the queen excluder is used as normal between the brood chamber and super. It is absolutely essential that all the joints between the supers, the clearer board and the crown board – or whatever is taking its place – are completely bee and wasp proof. If there is any small hole they can get in through then they will take the honey away and, especially in the case of bees, may start general robbing of any other colonies nearby. Bees and wasps have been known to take away the best part of thirty pounds of honey in a night. This becomes of greater importance as time passes and equipment and hives begin to suffer from general wear and tear. It is a good idea to carry pieces of foam rubber to stuff into any likely holes.

The bees will clear from the supers when using Porter bees escapes in one or two days depending on what is in the combs. If the comb is all sealed honey then the bees will clear very quickly and very thoroughly, but if there is much unsealed honey, and particularly if there is stored pollen present, then the bees will take longer to clear and usually there are a few left when the supers are removed. If using Canadian type escapes bees will clear faster, possibly as fast as 4 to 6 hours.

Supers not cleared

If there is brood in the supers then the bees will refuse to clear so if you look in the supers at the end of two days and find the bees still there in force it could be there is brood present or the escapes are blocked. In this case it is best to shake and brush the combs and take them away (see below for method). If however you do not feel competent to do this then take the supers off and examine the escapes. If these are blocked, possibly with drones which have accidentally got into the supers, remove the obstruction and put them back, then replace the supers for clearing. If the escapes are clear then examine the super combs for brood. If brood is found this may be because a queen has got up into the supers during a manipulation or because the excluder is damaged and the queen has found her way through. If possible check for the queen and if you find her run her back into the brood chamber. Check the queen excluder for any faults and realign the spacing if this is possible, or the queen will move back into the supers. The easiest thing to do is to put the super containing brood back over the queen excluder and then removing it later once all the brood has emerged. See 'Brood in Supers' page 107.

Removed supers

After clearing the supers should be taken to a bee-proof place and covered to keep the smell of honey in the box so that it does not attract bees should they find a way indoors. If the supers are taken into a room in the house, be sure to close the windows, including the small vents, or you may find the room full of bees within a fairly short time.

Extracting

Extracting should be done as soon as possible after the supers are taken from the bees while the honey is still warm from the hive. The warmer the honey is the lower its viscosity and the easier it flows from the cells. Extracting is made up of four separate processes: de-capping the combs, extracting (spinning the honey out) in the extractor, straining the extracted honey and putting into containers for storage. Most beginners have to extract in the kitchen and extracting can be a very messy process, making one very unpopular, unless good precautions are taken to prevent the sticky honey getting around. I would advise putting down a layer of polythene sheeting and having plenty of mopping-up cloths ready to clean up as the work proceeds.

Decapping comb

Sealed honey must have the wax cappings removed before the honey can be spun out of the cells in the extractor. Cutting off the cappings is called decapping or uncapping. I prefer to use the former as there is then no confusion between honey over which the cappings have been removed and honey which has never been capped. The amount and cost of equipment used to remove the cappings will depend upon the number of colonies involved, and hence the amount of comb and honey to deal with. The beginner only needs a large plastic bowl and a long knife such as a ham or bread knife (for hygiene reasons this knife should be kept for use with honey only). The bowl should have a piece of wood, or metal, across it. The point (the end of a screw will do) standing up in the centre of the wood is used to rest the side bar of the frame on. This should protrude high enough to allow the frame to be swivelled around on it without the lug catching on the cross bar. The job is done by taking the frames one at a time from the supers, resting the side bar on the spike and cutting off the cappings with about an eighth of an inch of the cell wall. If you are using Manley frames then the knife is rested on the top and bottom bars while cutting. Most people seem to find cutting in an upward direction gives them greatest control of the knife but it is necessary to make sure no part of your hand, on the top of the frame, is projecting on the knife side. If the knife slips, and it often does, you will cut yourself. If you can learn to cut downwards this prevents accident. The knife should be used with a sawing motion and the whole surface of the comb cut at one sweep if possible. Any hollows left with capping still on should be decapped with the point of the knife or a spoon. All the capping must be removed. The knife should be used dry and not dipped into hot water to aid in cutting; this has no advantage and only results in water being incorporated into the honey. The comb should be leaned over towards the knife so that as the cappings are cut off they fall into the bowl. If the frame is held so that the capping fall back onto the face of the comb they will stick back onto it and have to be scraped off, a messy procedure. Having decapped one side, the frame is spun around on the spike and the other side decapped. The frame is then put into the extractor so that any honey dribbling out will be caught. The decapping process provides most of the mess unless care is taken to ensure that sticky cappings are not flicked about all over the place where they may be trodden on and carried around on the shoes. When the bowl is full of cappings these should be put to drain in a piece of straining cloth tied to the top of a plastic bucket, this will get most of the honey out of the cappings for use. The cappings should be dealt with as described in 'Beeswax – recovery of' page 106.

Crystallised honey

If the bees have collected the honey in the combs from the brassica family of plants such as oilseed rape, then there is likely to be some crystallisation in the comb and this cannot be extracted in the normal way. Honey crystallised in the comb is best scraped from the central midrib of the comb using a table spoon or a heather scraper. This leaves the original wax foundation intact for the bees to rebuild the cells when it is put back into the colony. The scrapings, which are a mixture of wax and honey, should be warmed to a temperature of not more than about 93°F/34°C, until the honey is liquefied when it can be strained with the rest of the honey from the extractor, see below. Do not use higher temperatures or the honey may be tainted with the taste of propolis and wax.

Extractors

An extractor is a costly piece of equipment but the beginner can usually hire one at a very small fee from the local beekeeping association and can buy one when the honey coming in will pay for it. Extractors should be made from food grade plastic or stainless steel. The old galvanised ones, often given as a gift, should not be used as they may contaminate the honey. Extractors are of two main types: 'tangential' and 'radial'. In the former the combs are placed in the extractor at a tangent to the circular wall of the tank while in the latter type the frames are arranged like the spokes of a wheel. The tangential type has a wire cage, or screen, for the face of the comb to fit against, and be supported by, when the extractor is spinning and the force, which is at right angles to the face of the comb, is trying to throw the comb against the tank wall. Without this cage the combs will be torn from the frames and broken up. For this reason Manley type super frames cannot be used with a tangential extractor unless the screen is modified. The self-spacing side bars of the frame holds the face of the comb away from the screen and the comb is distorted or broken when the force is applied. Beekeepers with tangential extractors are advised to use ordinary, not self-spacing frames in their supers spaced with castellated runners. Manley frames are the best for use with radial extractors. The radial type only needs a slotted rotor, or cage, to anchor the frames in place as the force here is parallel with the face of the comb and does not easily break combs, unless very rapid acceleration is allowed. Once the rotor in the extractor is filled with combs it is spun and the honey thrown out of the comb by centrifugal force. The tangential type will only extract one side of the comb at a time while both sides are extracted at the same time using the radial type. Extractors may be turned by hand or by an electric motor. In the latter case it must be provided with a means of varying the speed of spin or many combs will be broken by rapid acceleration when the motor is switched on.

Extractors usually hold a quantity of honey under the rotor. The amount varies with the size of the tank. This weight of honey helps to hold the extractor steady and prevent it moving around, but it should be drained off before it reaches the bottom of the rotor or flows into the rotor bearings.

Extracting

Tangential extractor

When using a tangential extractor the decapped combs are placed in the cage, care being taken to balance the rotor as much as possible, and the rotor is set revolving slowly, either by hand or motor, and gradually speeded up until the honey can be heard pattering against the side of the tank. Keep revolving at the same speed, do not speed up for a couple of minutes. This will remove about 75% of the honey from the outside face of the comb. Stop the extractor and reverse the combs in the cage, so that the un-extracted faces of the combs are outwards, and then speed up once more until you hear the honey on the tank but in this case the speed or revolution should be speeded up to keep the sound of the honey against the tank at the same level until there is no further honey coming off from the outside face of the comb. The extractor is stopped once more and the combs reversed so that the first face of the combs can be fully extracted. It is absolutely necessary for these turnings of the combs to be made or the weight of the honey on the un-extracted side will break up the comb, particularly when combs are new; they get tougher with age.

Radial extractor

The decapped frames are put into the radial extractor and this is speeded up until the honey can be heard hitting the side of the tank, but with this type both faces of the combs are being extracted at once so the speed can be steadily increased, keeping the level of the sound constant, until the combs are empty. There is no need to stop to turn combs. Little damage will be done to combs, even new ones, unless too rapid acceleration is used.

This is an example of a radial extractor.

Straining

When the honey is taken from the extractor it should be strained to remove any large lumps of pollen and any cappings or pieces of comb which have found a way through. Probably, for the beginner, the best way is to hang a conical strainer on the tap and run it into a plastic honey bucket for storage. When larger amounts of honey are to be dealt with a honey tank is used and the honey poured through a nylon cloth strainer which is usually carried upon a built-in grid. These honey tanks usually come in three sizes, 25kg, 50kg and 100kg with detachable strainers. The main problem when straining is likely to occur in areas where plants such as oilseed rape are grown and there is a lot of incipient crystallisation in the honey which will rapidly clog the straining cloth. In such places a wire gauze strainer can be used so that the honey can be stirred to prevent the build up of crystals on the straining area.

Bucketing honey

Once the honey has been strained it has to be put into containers for storage or use. When there is only a small amount of honey it is tempting to put this straight into jars for the table but this has several disadvantages. The main problem is that many honeys, and particularly some of the very good ones, will set like concrete when they crystallise, making the result impossible to spread. This problem can be corrected by allowing the honey to crystallise and then to warm it and stir it carefully. Done properly the honey will re-set hard enough not to flow, even if the jar is inverted, but soft enough to spread easily. This process is best accomplished by putting the honey into tins, or plastic honey buckets, holding about 28lbs (12.7kg) or more and allowing the honey to crystallise. This has the advantage that a larger bulk of honey sets with a finer texture, keep better, is not so subject to fermentation as honey put into jars and can be stirred more easily and effectively after warming. If honey is stored in tins these must be heavily lacquered or the honey must be enclosed in a polythene liner or bag. Make sure the polythene is one suitable to hold food. Honey must not touch iron, of which most cans are made, as it will react and form a black 'goo' with a dreadful taste which will taint all the honey in the container. See 'Storing extracted honey' page 103 and 'Warming Honey' page 103 et seq. The second major problem of putting honey into jars as soon as it is extracted is that it will 'frost'.

Frosting of honey

At the end of the crystallisation air will come out of the honey and form a white marbled or foamy appearance on the side and the surface of the honey. There is also a change in the crystal formation which again helps to produce this whiteness. Frosting does no harm to the honey in any way but it does spoil its appearance and many people will think the honey 'is going off'. So if you are giving your honey to friends and certainly if you are selling it, frosting should be avoided. Again by putting the honey into buckets it allows the first crystallisation to take place and frosting to occur. When the honey is warmed and stirred the frosting is lost and honey does not frost again, unless it is heated above the temperatures given for bottling, see 'Granulated Honey' page 104.

Heather harvest

Heather honey because of its properties needs different treatment during extraction from any other honey. Because heather honey is a jelly it has to be pressed or as it is a thixotropic jelly it can be stirred and extracted in a tangential extractor.

Pressing

The heather press comes in many designs, two types are illustrated. They all work by squeezing the honey from the beeswax through linen scrim straining cloths directly into storage cans; it cannot be strained further. What is put into the press can vary. It can be the entire comb cut from the frame or the cells can be scraped from the foundation with the heather scraper. Both are wrapped in linen scrim for pressing. The latter method has the advantage that the foundation is in place for putting on the hives the following year. If this method is used the frames of foundation should be run through the extractor as 5 to 10% of the honey will be left on the foundation. Pressing is a slow business. When a large crop of heather honey has to be extracted, instead of pressing, the honey and wax mix scraped from the midrib can be spun out in a large purpose built type of spinner much like a domestic spin dryer (used only for this purpose).

Two types of heather presses.

Stirring – heather honey

The honey is stirred in the cells of the super comb using a tool called a 'Perfor Extractor'. This looks like a brush made with steel needles instead of bristles. The comb to be extracted is laid flat on a supporting surface and the needles of the 'brush' are pushed' in and out of the cells, piercing the central septum. After this the honey can be extracted in a conventional extractor. Large motorised equipment is available to do this without piercing the septum.

Cut comb honey

Probably the best way of using, or selling, heather honey is to cut the comb up into chunks and eat the lot. Heather honey lends itself to this method as it does not run or granulate, and there is always a ready sale for comb honey. Cut comb containers are available from equipment dealers.

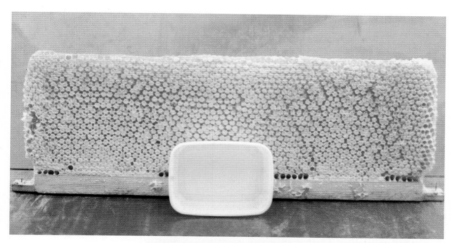

A full comb well sealed and ready to be cut to size to fit 8oz (272g) plastic containers.

Cut comb in container.

9. Honey

Honey composition

Honey is a natural product collected by the bees from the nectaries of many flowers and is therefore a very variable product. Honey's main constituents are sugars and water but it also contains small quantities of many other substances. An average analysis of its composition would be as follows:

Glucose	35%
Fructose	40%
Other Sugars	4%
Other Substances	3%
Water	18%

Glucose and fructose, also called dextrose and laevulose respectively, are two simple sugars which occur in all living things. The other sugars are of many different kinds and are mostly more complex sugars. Other substances include a vast assemblage of different ingredients: organic and amino acids, minerals, proteins, colloids and plant essential oils.

Properties of honey

Understanding the properties of honey is of extreme importance when one is handling honey during extraction, storage and putting into containers for use or sale, because the aim of the beekeeper is to provide for the table a commodity which has been as little altered as possible from the substance which was stored by the bee. The most important properties are flavour, colour, water content, hygroscopicity, viscosity, crystallisation and fermentation.

Flavour

The basic flavour of honey is sweetness because of the amount of sugar it contains. The sweetness is to some extent counteracted by the acidity of honey, coming from its acid content, which is quite high. The flavour is also enriched by the essential oils of plants, including the substances which provide their scent and colour, these giving honeys from different plants a unique flavour. On average this type of flavour is least present in light honeys and increases in strength with the darkness of the honey. Little can be done to influence the flavour of the honey your bees will collect. The only way to affect it is to take the colonies to crops which will provide a special flavour. The flavour can easily be ruined by overheating or poor storage methods. Unfortunately the fine flavours of all honeys are lost during crystallisation and nothing can be done about it. Only a beekeeper knows the glorious taste of a piece of wild comb, filled with good quality honey, and eaten warm from the colony. This is an incomparable delicacy.

Colour

Colour is again dependent upon the plants from which the nectar has been collected. It can vary from water white honey of borage and red clover, through the darkening yellows of clover and dandelion to the dark amber of ling honey and the deep port wine colour of bell heather. Overheating can darken honey as well as spoiling the flavour.

Water content

Water content is of extreme importance because it affects the keeping qualities of the honey and whether it is legally fit for sale. The beekeeper controls the water content of the extracted honey by only taking it from the bees when it is ripe and ready. See 'When to take honey' page 89. Extracted honey with a water content exceeding 20% is not legally saleable and above 21% will ferment quite quickly and be fit only for using in cooking. see 'Fermentation' overleaf.

Hygroscopicity

Hygroscopicity is the property of absorbing water from the air and thus increasing the water content of the honey. Because of the large amount of sugar and the small amount of water present in honey the latter is extremely hygroscopic and care should be taken at all times not to expose the honey to air any more than is necessary. This applies to both clear and crystallised honey. During storage or use the honey should be kept in air tight containers to prevent deterioration, and ultimately fermentation. See 'Fermentation' overleaf.

Viscosity

Viscosity is the property of honey which causes it to resist flow. It is basically caused by the large amount of sugar in the small amount of water and is increased by the amount of proteins and colloids present in the sample of honey. From the practical point of view the viscosity is only important in so far as it hinders the straining of honey. The greater the viscosity the more difficult it is to get the honey through the straining cloth. Fortunately the viscosity can be lowered sufficiently to make straining easy by warming the honey to about 93°F/34°C. Its other important effect is to affect the size of the crystals formed when honey crystallises. A very viscous honey will often form quite distinct crystals and be gravelly in texture, which is appreciated by some but not by the majority of people.

Heather honey

This honey, which is collected from the ling, *Calluna vulgaris*, has in it a greater amount of protein and colloids than most other honeys and these are sufficient to turn the honey into a jelly. It is a 'thixotropic' gel that will turn into a fluid if stirred and return to a gel if left to stand. This means it needs special handling during extraction and bottling. Absolutely pure heather honey never granulates but if it is collected with the nectars of other plants then these will granulate. Small amounts will appear as crystal aggregations distributed within the bulk of the heather gel ranging from the size of a mustard seed to green pea size dependant on the amount of other honeys present. There appears to be little mixing. The presence of about 40% other nectars will induce a complete soft granulation.

Crystallisation

Honey before it has crystallised is usually termed either 'liquid', 'clear' or 'run' honey while after crystallisation it is called either 'crystallised', 'granulated' or 'set' honey. Crystallisation is caused by the large amount of sugars dissolved in the small amount of water making the honey highly supersaturated. Given time almost all honeys will crystallise. Some honey which contain a high percentage of glucose, such as honey from oilseed rape and raspberry, will crystallise within a couple of days while others being low in glucose and high in fructose, such as false acacia *Robinia pseudoacacia* honey, will rarely set at all. The vast majority however granulate within three months of extracting. Rapid crystallisation will usually produce a granulated honey with a very fine texture while if the honey is slow to crystallise then the texture may be quite gravelly. This coarse granulation can be avoided by using the process of 'Seeding', see page 117, during handling. Granulated honey is even more hygroscopic than clear honey and must also always be stored in airtight containers.

Fermentation

Fermentation is the process which occurs when sugars are digested by yeasts as they grow and reproduce. Fermentation in honey is caused by yeasts which normally live in the nectaries of flowers and are collected by the bees and brought back to the hive in the nectar. When the bees lower the water content the high concentration of sugar kills most of the yeasts. Some survive but are prevented from reproducing. Therefore in clear honey with a water content of 20% or less fermentation cannot occur. Above this water content, other conditions being right, the honey will ferment. Once honey has crystallised and much of the glucose has been turned back to solid crystals the liquid between these crystals will have a water content of about 4 to 6% higher than the original honey, because 90% of the original water will now contain the smaller quantity of sugar still in solution – the other 10% of water will have combined with the crystals. Granulated honey is therefore always open to fermentation unless it is stored below 50°F/10°C. Yeasts of this type can reproduce at temperatures of between 50°F to 80°F (10°C to 27°C). Honey stored below the lower temperature is safe from fermentation at all times. Fermentation occurs in two main ways. Firstly, where honey has been allowed to take up water from air, there will be a thin fluid layer of honey on the surface with a very winy smell. Secondly, the more usual way is for the surface of the honey to remain dry but heave into hills and dales, similar to a working baker's dough, and with a strong smell of fermentation. The honey is still quite alright to eat. It can be heated to about 150°F/65.5°C for ten minutes or so and used for cooking but cannot be sold unless labelled as 'Baker's Honey'. Honey stored in buckets should be filled to exclude as much air as possible and the lid tightly secured.

Storing extracted honey

Honey must be stored in airtight containers in a cool place, preferably below 50°F/10°C, which is not too difficult in Britain except perhaps in July. Honeys which set very hard keep longer than soft honeys and all honeys keep better the larger the container, and hence the recommendation to store honey in bulk. Most honeys stored in the usual 28 lb (12.7 kg) plastic honey bucket can be kept for two years, if necessary, before fermentation starts.

Warming honey

When honey from stores is required for use it has to be warmed sufficiently to get it out of the buckets and into smaller suitable containers, usually the standard 12oz or 1lb honey jars. The temperatures used in warming honey are fairly critical as it is easy to overheat honey and caramelise the outside layers of the mass, as heat penetrates very slowly and mostly by conduction. Usually a simple method is found of warming in one's first couple of years of beekeeping using slow burning stoves, airing cupboards or even the greenhouse, but before long a warming box will have to be made. The simplest way is to get hold of a small broken down household fridge which will provide a very nice insulated box. A hole to take an electric cable is bored through the wall and a light fitting is fixed to the floor, a forty to sixty watt light bulb (not long life) will provide all the heat required. The bulb should be connected through a thermostat fixed near the top of the fridge compartment, with a second hole to allow a thermometer to be pushed through near the top of the fridge to check and calibrate the thermostat. If the original position of the bottom shelf is right to take a honey bucket all is well. Otherwise a frame will have to be made to provide a shelf at the right height to get a bucket in under the top of the compartment. It is best if something, like a thin slate or piece of block board, is placed above the light to prevent the heat concentrating above the light bulb on the bottom of the containers. The temperatures used will vary as to whether the honey is to be bottled as clear or granulated honey and are set out overleaf.

Granulated honey

When bottling granulated honey it should be warmed just sufficiently to get it out of the bucket into the bottles, and the crystals should hardly be melted at all. The warming temperature used when the honey is in 28 lb buckets, is 90°F/32°C applied for one to two days, depending on the hardness of the honey being warmed. If you are doing more than one bucket at a time then the hard honey should be put in first for a day and the softer later. It should be realised that no matter how good your warming box is there will always be hot and cold spots. It is therefore best to turn single buckets once a day or to swop around buckets if there is more than one. After two days very hard honey will still not run out if the open can is inverted but if a wooden spoon is pushed down to the bottom of the bucket and pulled up and down vertically a couple of times the crystal structure of the honey will be broken and it will flow quite readily. The honey texture should be like smooth porridge. Single buckets of honey can be ladled into jars or, better, poured into a large jug or the honey tank and from that into the jars. If more than one bucket is being done at one time then they should be emptied into the honey tank used at extracting time and bottled from that through the tap. In the latter case the honey from the different buckets should be gently stirred together to give a blended honey of one colour. Little mixing occurs if this is not done resulting in striped honey in the jars.

Clear honey

Bottling clear honey is a much greater problem and my advice is to use, or sell, your honey granulated unless you have a great deal of it. In this case all the honey crystals have to be melted back into solution. This is done by heating in the warming box at 125°F/52°C, for up to 2 days or until it is all liquid. The honey should then be poured through a straining cloth into another bucket or the tank to take out any small lumps of pollen or wax which may have been left in the set honey with no disadvantage but which, in clear honey, will cause it to appear cloudy or show up as floating specks. Having got the honey into jars if left as it is it will crystallise again, often within a few days, and this time with very coarse crystals. It is necessary therefore to heat it again to keep it clear for a reasonable time. This second heating is done by placing the full jars in a water bath on a trivet at 145°F/62°C, for one hour. The lids should be on the jars screwed down and the water not allowed to reach above the shoulder of the jar. The water should not touch the bottom of the lid. After removal from the water bath the jars of honey should be cooled as rapidly as possible, placed in a cool draft if possible and not bunched tightly together. The water bath is necessary as air heating is not sufficiently swift to provide a shelf life without darkening the honey.

Heather honey

This unique honey can be easily spoiled by overheating. To get it from bucket to bottles it needs warming at 115°F/46°C, for about two days, stirring and pouring into the honey tank for bottling. It will reset in a few days and be ready for use or sale.

Selling honey

Should you think of selling some of your honey then you must find out what legislation is in force at the moment. At the time of writing, the Honey (England) Regulations 2003 have been amended by the Honey (Amendment) (England) Regulations 2005, No. 1920. The regulations for Scotland and Wales are similar. The following are brief notes on the position in the United Kingdom, and to some extent the EC.

Previously extracted honey could only be sold in specified weights those usually being ½ lb (227g) and 1 lb (454g). From April 2009 a new Directive (2007/45/EC) came into force, which led to the deregulation of all specified weights for honey and some other products. However the container must be correctly labelled as described below. It is not legal to sell a pot of honey without a label even from your own house. The label must have the following information displayed on it: 'Honey' which can be extended by words such as 'Pure', and words to indicate the geographical origins such as 'Essex' – providing the statement of origin is correct, words to indicate the floral origin as long as these are true statements. In most areas the honey is of a mixed origin so the use of a floral specification would be incorrect and it would even be incorrect to use the picture of a flower on the label unless the honey came from that flower. The exception in Britain is the use of 'Heather Honey' for the honey produced from ling *Calluna vulgaris*, which is so unique that it is easily recognised.

The label must indicate the country of origin: e.g. 'Country of origin: UK'. The label must show the name and address of the producer (or packer or retailer) and the weight of honey it contains, the latter should be in grams (pounds being optional), using the letter 'g', these figures being 4 mm in height for 1 lb, 12oz and ½ lb jars, grams before pounds and sited in close proximity. Standard labels can be obtained from the appliance firms, or some County Associations, and the seller's name can be written in or applied with an indelible stamp, but must be shown on the main label. In addition there should be a 'lot number' prefaced by the letter 'L' and a 'Best Before' date. If the Best Before date uses the numbers for day/month/year, the lot number can be omitted. A record of the date on which the honey was bottled should be kept and this should tie in with the lot number on the label. There is no recommended time limit for the 'Best Before', but a sensible limit would be one year to eighteen months from the date of bottling.

10. Techniques

This chapter contains a number of beekeeping techniques which the beekeeper will use when required, and details of some equipment and substances for use by the colonies. It is arranged in alphabetical order.

Ethanoic (acetic) acid fumigation

Ethanoic acid is used to fumigate used comb and will kill Nosema spores and wax moth eggs and adults. This is an acid so take precautions. Use an overall, butyl rubber or neoprene gloves, eye protection and a face mask. If the acid contacts the skin wash off under a tap. Use this acid diluted to 80% (purchased from beekeeping suppliers). Do this outdoors and protect any concrete. Protect metal ends and metal runners with a coat of Vaseline. Stack the boxes on a solid floor with entrance closed. For each box place an absorbent pad over the top bars and add about 100ml of acid, then the next box and so on. The acid fumes are heavier than air. Close the stack ensuring it is bee tight, as despite the strong smell bees will rob the boxes. Leave for one week (longer if the weather is cold), then air the boxes using screens to stop bee access.

Bailey comb change

This method was advocated by Dr Bailey of Rothamstead for managing colonies with Nosema. Having prepared a new brood chamber with new or fumigated drawn brood comb in preference to foundation as this colony will not be strong enough to draw comb. The number of combs depends on size of ailing colony, use a dummy board to contain brood nest. Find queen and place her in the centre of the frames of the new brood box. Remove all comb not containing brood from old brood chamber and use a dummy board. Fumigate the removed comb using ethanoic acid if the comb is to be reused or destroy by burning. Add the queen excluder then shallow frame called an 'eke' about 1–2″ deep with small flight hole facing in the same direction as old entrance. Add new brood chamber containing queen and new comb. Close old entrance. Feed with syrup in a frame feeder or contact feeder. One week later if queen has moved onto a new comb move old comb downstairs. In three weeks all brood in lower brood chamber have emerged, remove the bottom box and rearrange the new box onto clean floor. Add more drawn frames as colony expands. Feed as necessary.

Beeswax – recovery of

Beeswax is a valuable commodity and every bit should be saved and recovered. There are two main sources, old worn out comb and cappings. The latter will be dealt with under 'Cappings - disposal of' page 108. It is very difficult to recover on a small scale all the beeswax from old comb but a large proportion can be melted out using a solar wax melter. The solar is an insulated box with a double glazed lid. Inside the box is a raised metal tray running into a removable metal catching vessel. The metal must not be iron as this reacts with the wax and turns it dark brown. The whole box is sloped at an angle of about 40° from the horizontal and placed in the sun facing due south. A box of this sort will reach up to about 190°F/88°C, so any insulation used in its construction should be able to take this sort of temperature. Beeswax melts at about 145°F/62°C, so it will be recovered from comb almost any time the sun is out for a few hours. Large boxes are more efficient than small ones and I prefer an extractor which allows one to put in the frame as well as the comb, so that spores and eggs on the frame are destroyed. If the comb or frame of comb, is wrapped in a muslin bag, or old nylon stocking, this strains the wax at the time of recovery and a cleaner sample of wax results. Not only old comb but all the pieces of brace comb, torn out queen cells and bits of drone comb should be collected and placed in the extractor for wax recovery. These should certainly not be thrown about in the apiary. Good clean beeswax can be turned into foundation and used again in the colonies, or if you have more than you require it can be easily turned into candles, furniture polish, face and hand creams, or used to wax threads or rub on hot pottery to waterproof the vessel.

Brood in supers

A few scattered drone pupae in the supers, recognised by the tall drone capping are of no importance, probably evidence of the odd laying worker which seem to turn up at times. It is the large amount of brood in the supers that we are worrying about under this heading. This is evidence that a queen has got into the supers and her presence there can be caused by a variety of circumstances. In the following, I am assuming that the supers are filled with worker comb.

1. There is worker brood in the combs on both sides of the excluder. In this case the most likely cause of the problem is a faulty excluder or a very small queen. If the colony is large it is unlikely to be the latter but replacing the excluder with another will usually cure the problem. When the new excluder is put on, the queen should be found and placed in the brood chamber, or the excluder can be put in and four or five days later the colony can be examined and the box containing eggs will contain the queen. If it's the brood chamber all is well but if it's the super then she must be found and put down. If eggs are still found on both sides of the excluder then you have a small queen or a lot of excluders which need replacing.

2. There is no young brood in the brood chamber but plenty, in the super. In this case the queen is in the super and must be found and put back into the brood chamber. The probability is that when you were manipulating you missed seeing the queen on the

excluder and she was on the super side of it when it was put back on. You can calculate when this happened by looking at the age of the youngest brood in the brood chamber.

3. There is plenty of normal brood in the brood chamber but the brood in the super is capped drone and there is plenty of it. A virgin queen has got trapped in the super during swarming or supersedure and has been unable to go out to mate but has come into lay, and as the eggs cannot be fertilised, producing nothing but drone brood. The queen in the super must be found and killed. The drone brood can be killed by running a hive tool over the cappings, or if left then be careful to release the drones or they will block the excluder when they die. If you should use drone foundation or comb in the supers you may miss this last situation, and on putting the queen down from the super, may kill the mated queen. If a colony is found with no brood but they are handling normally and are not 'touchy', always look into the super to see if the queen is up there before getting all excited and thinking the colony is queenless. If she is not in the super put a test comb in the brood chamber, see 'Test comb' page 74.

Cappings – disposal of

During extraction of the honey crop the cappings are strained of most of the honey but are still a sticky mass at the end. There are a couple of ways of dealing with them.

1. They can be washed in water and strained, the resulting honey water can be made up to the required gravity and turned into mead.

2. The sticky cappings can be placed in a Miller feeder with the cover strip removed so that the bees can come up and clean the cappings. The bees will turn the cappings over and clear up every bit of honey. Having cleaned the cappings of the honey they can be placed in muslin into the solar wax extractor or placed in a vessel over a small amount of water and heated until the wax is all melted, the whole can be left to cool and the block of wax removed when it has shrunk away from the vessel. The underside of the block will be covered with a brownish substance, slum gum, which is normal and can be scraped off. Good colour cappings can be used for making candles.

Clearer board construction

The ordinary clearer board is the crown board with two Porter escapes in it, see 'Crown Boards' page 27. An easier method of construction is to leave a gap of about ½ inch (12 mm) between the two ½ inch (12 mm) thick boards, and covering this gap on the underside of the board with wire gauze, or perforated zinc, leaving an exit hole each end and covering the top of the gap with sheet metal leaving an entrance hole in the middle about 1 inch (25.4 mm) long.

Comb – fumigation of comb using acetic acid

Acetic acid 80% is used to fumigate comb. Acetic acid will destroy Nosema and Chalk Brood spores and also the adult wax moth and their eggs. Stack the box of frames to be fumigated on a hive floor with a solid block in the entrance. Pour about 100 ml of acetic acid onto an absorbent pad placed on the top bars of the frames. Acetic acid fumes are denser than air and will sink through the box. Repeat this process with all the boxes needing fumigated. The top box should be closed off with a solid cover board and roof. The stack should be made bee tight, with all holes closed off with pieces of foam, as bees will rob the stack despite the strong smell of vinegar. The treatment takes one week in warm weather and longer if it is cold. The pad can be left on all winter, but the stack should be checked regularly for the presence of wax moth. The boxes should be aired before being replaced on a live colony. Note that when using acetic acid you should wear a face mask, goggles and gloves. Acetic acid will attack metal so protect metal ends and metal runners in the boxes with a covering of Vaseline. Acetic acid will also attack concrete.

Clipping the queen's wings

The best time to clip queens' wings is when they are introduced or when you happen to see them at the start of the season. If they are not found by the end of April, before the swarming season is likely to start, then they must be searched for and clipped. The beekeeper should have a pair of small nail scissors ready. It's a good idea to have them in a pocket attached by a piece of string to the overall pocket. The method is to look through the colony, using as little smoke as possible, and find the queen – see 'Finding the queen' page 67. The frame with the queen on it is laid down gently on the top of the brood chamber or even better into an upturned roof so that she cannot run back into the hive, and, keeping an eye on the queen to see she does not walk off the face of the comb, gloves are removed and the scissors fetched out of the pocket and left dangling on the string. For right-handed beekeepers the queen is picked up by the wings with the right hand and immediately placed on the ball of the thumb of the left hand. The first two fingers of the left hand are placed over her head and shoulders and she is then allowed to creep in further under the fingers as her wings are released. The scissors are then slipped under the two wings. Watch that she does not put up a leg into the scissors. Then snip off about two thirds of one or both wings. I prefer this method of holding the queen rather than holding her between the fore finger and thumb as the queens struggle less and one is less likely to drop them. Once the queen is clipped she should be taken back between the fingers of the right hand and then released onto brood on the frame she was found on. If the bees begin to snap at her do not interfere but quickly pick up the frame and return it into its place in the brood chamber. Normally the queen will then be quite safe.

Marking queens

Many people mark their queens which makes them easier to see when they are wanted. Marking can be done in several ways. The simplest is to use paint and place a dab on the queen's thorax. The paint must not have an amyl acetate base. This is too near the pheromone which causes bees to sting and the queen will be killed. It is probably best to buy paint from a beekeeping equipment firm as this will be well-tried. Small plastic numbered discs can be obtained which can be stuck onto the queen's thorax with the glue provided. To do the job find the queen and pick her up as described for clipping. When the queen is safely caught by the fingers of the left hand slide the two fingers over her back and down, one on each side, trapping her legs on the ball of the thumb. Mark her and wait a couple of moments to let the paint or glue dry and then replace her as before. When using paint take care not to drown the queen in paint, place a drop on your glove first to take up any excess paint, then a small spot onto the queen's thorax. See 'Clipping the Queen' page 109.

Moving colonies

Forager bees learn to accurately orientate to their hive using both sun compass and their memory of the local landmarks over their whole flight range of about 1½ miles/2½ km. When colonies are moved this has to be taken into account. The old rule of thumb is that 'colonies must be shifted under 3 feet or over 3 miles'. If shifted more than 3 feet many bees will not find their hive, although their ability to do so varies with different strains of honey bee. Shifted under 3 miles some of the bees will, on flying out, reach their old flight range and return to their previous site. When colonies are being moved they have to be shut in but this must be done properly or the colony may be killed or severely damaged. Confined bees build up very considerable heat, sufficient to lower the strength of the beeswax in the combs so that these will collapse and kill or half cook the bees themselves. To prevent this happening the colony should have its crown board replaced with a 'travelling screen', which allows the heat to rise rapidly out of the hive. The travelling screen is made of wire gauze with an inch square wooden framing all round and one piece of wood across the centre to support the wire. This framed screen can be fixed firmly into position with four screws into the top box on the hive. The floor, brood chamber and supers are securely fastened together using long staples or triangular metal plates. If staples are used they should be driven in at an angle, or the boxes may twist and lift slightly when being carried. Alternately all the hive parts can be fixed together with straps (obtained from most bee suppliers). It is best to use two straps in parallel. If used at right angles as in tying a parcel, the hive parts can swivel one on the other and release bees.

The bees are shut into the hive by stuffing a 1¼ inch/32 mm square strip of plastic foam into the entrance in place of the entrance block. The hive entrance must not be shut using gauze or perforated zinc because the bees will try to get through the holes and many will be injured in the scramble.

The usual procedure is for the screen to be put on during the day, the crown board and roof put back on above it. If this is not done many bees will be attracted by smell coming through the screen and alight on the outside of the screen and will be a nuisance when loading the hives and lost later on during the journey. The plastic foam is stuffed into half the entrance, and left with the end sticking out at right angles. When the bees stop flying the rest of the foam is pushed in, the crown board and roof removed, so that the screen is open to the sky. Hives should be loaded with the frames parallel to the direction of movement of the vehicle. If they are to be in transit for some while then try to move at night. Spray the screen with a clean water spray. Should the colonies begin to roar spray the screen again generously with the water spray. This sprayer should not have been used previously for insecticides. On arrival at the new site the hives are placed out into position, and their crown boards and roofs leaned against them and then, starting at the furthest colony from the vehicle, remove the foam from the entrance and quickly place the crown board and roof on the hive. Let the bees out before the crown board is put on or the sudden shutting off of the top ventilation may irreparably damage the colony by overheating. It is obvious that if you can have two people doing this, one opening and the other dealing with crown boards and roofs, the job can be done quicker and with less fuss. It is best to leave the screens on at this time and remove them when the colonies are next examined, by which time they will have settled down to work and cause no problem. If the move is only a temporary one and the colonies are soon to be returned to their old site it is necessary to number the old site and the colonies so that they go back in the same position as before, because many of the bees will remember their old positions for up to three weeks and if in the wrong place the resulting mix up of bees may cause fighting and loss of queens.

Nuclei making

Nuclei have many uses in the management of honey bee colonies. Three major uses are to increase the number of colonies kept, to mate young queens in spring and summer, and to have queens available to act as replacements for failing queens during the season and also in the following spring. Here we will deal with making increase, the other two uses being included under 'Queen Replacements – production of' page 116. Nuclei can be made under two different sets of circumstance; one where they are to stay in the same apiary – when any flying bees incorporated into a nucleus will fly home to their original colony – and two; where the nuclei are to be taken to an apiary over 3 miles / 5 km away, in which case any bees put into the nucleus will have to stay as they will find themselves on strange ground. Where possible nuclei should be moved away to a strange site as this makes the process much easier to control. Before one starts to make a nucleus it is necessary to ensure that there will be a new laying queen available or a ripe capped queen cell. On no account should the nucleus be allowed to make a queen cell of their own; this is the way to produce a very small poor queen who will rarely be of any economic value.

It is usual to make a nucleus by taking frames of brood and bees from a strong colony. Having decided on which colony to use, approach the colony quietly and open it gently to find the queen. Once found, the queen is placed in a safe place on the comb she is on into a nuc box and covered, or into a match box with a few workers to look after her. If the latter, it is best to place the match box in the entrance of the hive so that if you should forget to put the queen back the bees will gnaw a hole in the box and let her out. This is necessary to ensure the queen is safe and that it is impossible to put her into the nucleus. Do not cut out this part of the procedure as, it does not matter how skilled you are, you will miss the queen sometimes. If she ends up in the nucleus the colony will be set back and the nucleus will have killed the new young queen, or queen cell, you have put into it.

From now on it will depend upon whether the nucleus is to remain in the apiary or be moved away. If it is to stay in the apiary then three frames of sealed hatching brood and a frame of sealed stores, together with the bees that are on them, are placed in the new brood chamber, and a further four frames of bees are shaken in to help look after them. Before putting the frames of brood or shaking the bees from the other frames into the nucleus, they should be given a gentle shake over the original hive as this will dislodge a lot of the old bees who would fly home if incorporated in the nucleus. The nucleus can now be put on its permanent site. If it is to be given a mated laying queen, she can be put in a Butler cage right away, see page 115, and the frames of foundation put in and the crown board and roof put on. The nucleus can now be treated as described in Chapter 4.

If the nucleus is to be given a queen cell, it is made up with less frames of brood. It should have one frame of hatching sealed brood and one containing young unsealed brood and eggs. Young bees should also be added as described above. Don't forget the frame of sealed stores. This will mean that there will be sealed brood in the hive for three weeks and there will still be some left when the new queen begins to lay. It helps to stabilise the colony during the period when it only has a virgin queen. The nucleus must be left for two days to become aware of its queenlessness. If the queen cell is put in straight away then it will be torn apart and the queen pupae killed. After two days the nucleus is opened, any queen cells made by the nucleus destroyed and the ripe queen cell added by gently pressing it slightly into the face of the comb above brood. There is no advantage in giving this nucleus frames of foundation until the queen has emerged and started to lay as bees in this state will not draw foundation very well if at all. In neither case should the nucleus be fed as the older bees returning home may come back to fetch the syrup and rob out the nucleus at the same time. Feeding, if necessary, should be done a couple of days later.

The process is slightly different for the nucleus which is to be taken away to a distant site. Assuming the queen is safely packed up, four frames of hatching sealed brood are placed in the new brood chamber, or nuc box, and a further two frames of bees are shaken in. There is no need to pre-shake the frames this time as any old bees will not be able to go home anyway. The hive is closed and taken away to its new site. In place on the new site the nucleus is opened and the new queen introduced, the foundation put in and a feeder put on with about a gallon of heavy syrup.

If the nucleus is to be given a ripe queen cell then instead of four frames of sealed brood it would be better to have three sealed and one with young larvae and eggs. The nucleus should be given about a half gallon of syrup. It must be kept queenless for two days, any attempt to make its own queen cells removed and the chosen queen cell is put in. In this case do not use foundation. If they are short of stores another half gallon of sugar syrup can be given. As soon as the queen is laying it can be given foundation and another feed. Both types of nucleus can have normal treatment from then on, see Chapter 4, page 50.

Having made your nucleus do not forget to replace the old queen into her colony. If she has been in a match box the best way is to take out a frame of brood and release her onto it and then put it back into the hive.

In all cases when nuclei are made up it is worth stuffing the entrance of the nuc box or hive with fresh green grass. This will wither fairly quickly, and allow the bees to work their way through, but it does stop the initial 'rush out' of bees in the morning and emphasise the fact that something out of the ordinary has happened.

Out apiaries

Some beekeepers may need to keep their bees away from home in an 'out apiary'. Certainly those who get enthusiastic and want to keep a number of colonies will have to keep their bees on more than one site. Few permanent sites will be able to support more than a dozen to fifteen colonies. For the convenience of neighbours suburban garden apiaries should not have more than four to six colonies. Many farmers, although not beekeepers themselves, appreciate that bees are good for pollination so are very happy to have bees on their land. They will allow beekeepers to occupy a small corner of their land out of their way. Out-apiaries should be established in areas where there are plenty of honey bee forage plants. There should be a hard access path suitable for a vehicle so that a lot of heavy carrying is avoided. The site should be secluded to avoid vandalism and to provide wind cover for the colonies. Apiaries should not be at the bottom of a frost pocket nor under heavy tree canopy which would encourage dampness. The layout of the apiary should reduce drifting to a minimum. Hives placed in a straight line will cause bees to enter neighbouring hives, mistaking them for their own. The hives placed in a circle, or as near that as possible, is the best for this purpose. The site may need fencing if there is livestock around. A sturdy fence is needed, but do not use barbed wire unless you have the farmer's permission because some will not allow it on their land.

Placing bees on the heather.

Placing bees on a field of oilseed rape.

Queen introduction

Introduction is the process of successfully putting a new queen into a colony. It is one of the most difficult techniques in beekeeping. The colony into which a new queen is to be introduced must be queenless. At the beginning of the season, while the colonies are still small, or at the end of the season, when the honey has been taken off and the colonies settled down for winter, the new queen can be introduced directly into the colony as soon as it has been made queenless. The usual method is to examine the colony, find the queen and remove her, and then the new laying queen is put into the colony in a Butler cage. The Butler cage is made of wire mesh bent into a square tube about 3½ inches (8-9 mm) long and ¾ x ½ inch (19 x 12.7 mm) cross section. The mesh should have holes about ⅛ inch (3 mm) square. One end of the tube is filled with a block of wood and the other left open. The queen is placed in the cage completely on her own and the open end of the tube covered with a small sheet of newspaper held in place with an elastic band. The cage is then inserted between two central frames with the papered end near brood. The bees will chew their way through the paper and let the queen out in the next few hours. It is a good idea to drive a panel pin into the wood plug, leaving it projecting, so that it sticks into the comb and prevents the cage slipping to the floor. The colony should not be touched for at least a week. Should the queen be still in the cage, but alive, when you next examine the colony do not tip her out of the cage into the hive because she will surely be killed. It is better to go over to the nucleus method, as set out below, placing a new piece of paper over the end of the Butler cage.

During the rest of the year it is best to introduce the new queen to a nucleus first and then to the main colony later. The colony to be re-queened is opened and the queen caged. A nucleus is made with one frame of brood, one frame of stores and a dummy frame on each side, a further four frames of bees are shaken in to keep everything warm and the new queen is placed between the frames in a Butler cage. The nucleus is placed by the side of the colony, the opening facing in the same direction as the parent hive, and left undisturbed for a week. At the next visit the nucleus is opened the cage and dummies are taken out, and the combs examined to ensure the young queen is laying. It is best to find the queen and make sure she has not been injured. The two frames are then placed in the centre of the nuc box and left open until required. The light will drive the queen into the dark between the frames. The main colony is now opened, the old queen found and removed, and the frames pushed up to one end leaving space for the nucleus. The two frames of the nucleus are picked up together and placed in the main colony, which is reassembled and left alone for a week. It is a good idea to spray the nucleus and colony with water, without separating the frames, to stop the bees on the top bars, and outer faces of the combs, from running about for a moment or two. In my experience this method loses fewer queens than any other method I know.

As an alternative add a few empty frames or foundation to the nucleus, now placed into a normal brood box, which can then be united to the main colony using the newspaper method. See 'Uniting Colonies' page 120.

Queen replacements – production of

Queen rearing is one of the fascinating parts of beekeeping but is rather outside the scope of this book. I would like to give a couple of simple methods of producing a few queens for the beekeeper with up to about half a dozen, or less, hives. The easiest way is to make up a nucleus from a colony that is swarming and include one good queen cell on one of the combs. The nucleus can be put into a nuc box and left for the queen to emerge, mate and start laying. The disadvantage is of course that you have no control over which colony shall provide the cell, and when, or even if, it will make cells. In my experience the wonderful colony you wish to multiply makes no queen cells whereas the ones you would like to lose are at it every year.

A more certain method is to build up your best colony onto two brood chambers, plus as many supers as necessary, until about June when the two brood chambers are split. The queen should be left in the bottom one then an excluder placed on it, then at least two supers, a second excluder and then on top the other brood chambers full of brood. Usually the colony will make a few queen cells in the top box and these can be taken to make up several mating nuclei. The disadvantage with this method is that you have no control over the number of queen cells made or the age of the larvae taken to produce them. If larvae are taken too old then poor queens result. This method is a modification of the Demaree method of swarm control.

Finally if a little more trouble is taken with the job better queens will be the consistent result. Again a colony is built up onto two brood chambers and supers. In this case it need not be the colony you wish to raise queens from. Around June the colony is opened and the queen is found and a nucleus is made to contain her. Two or three frames of brood and bees plus a couple of frames of foundation will suffice. A week later the colony is opened and the queen cells all destroyed and a frame of brood from the colony you wish to breed from is added for the bees to make queen cells on. This frame should have just hatched larvae and eggs. Eight days later the cells will be ready to put out into mating nuclei. As soon as the queen cells have been removed the queen can be reunited with her colony, see 'Queen Introduction', page 115. Queen cells must be cut out carefully with a piece of comb at their top. If they are damaged in any way the bees will quickly destroy them. They can be fastened into the mating nuclei by pressing the piece of comb at their top into the face of the comb in the nuc. Queen cells should always be positioned so that the queens emerge onto brood. Some of the mating nuclei can be built up into five or six frames and overwintered thus providing an early source of young queens in the spring.

Seeding honey

If you know your honey is going to granulate coarsely, because from past experience you know it always does, then the honey can be 'seeded' at extraction time. At least 1lb/500g, of finely granulated honey should be added to each 28lb/12.7kg, honey bucket before it is filled with the new honey. A gentle stir will help. The finely granulated seed honey can be from last year's seeded crop. It should be gently warmed in the warming box as one would to bottle granulated honey, see page 104, so that it is soft and will mix with the new honey. If you never have a fine seed it is worth buying a jar of fine honey and use this to do your first seeding. If only some of your honey is coarse, and you have no idea which, then as you come to use the honey select the best and put it aside for using as seed for next year. The granulated coarse honey should be heated in the warming box as for bottling clear honey. The seed is warmed to soften it, and then the above proportion can be added to each bucket. Care must be taken that the temperature of the honey to be seeded is not greater than about 80°F/26.6°C, or the seed crystals will be melted when stirred in and the whole exercise will be a failure.

Shook swarm

A shook swarm is the current preferred management of a colony suffering from a mild to medium infection of EFB, as mentioned in Chapter 7. The Seasonal Bees Officer will assist you with this manoeuvre. You will need to have to hand a spare disinfected brood box, floor, crown board, up to eleven frames of foundation and a rapid feeder, plus at least two gallons of heavy sugar syrup. The infected colony is moved to one side and the new brood box, floor and some of frames placed on the original site. A queen excluder is placed above the floor to stop the bees absconding. The queen should be found and placed in a cage such as a Butler cage and placed over the queen excluder. The bees will release her. The bees are gently shaken and then brushed off each comb into the new hive. The frames of brood are destroyed by burning and the old brood chamber, floor and crown board scorched with a blow lamp. Gloves, smoker, hive tool, and queen excluder are scrubbed with a solution of washing soda. The colony is fed with the syrup a couple of days later to encourage them to draw the foundation. The delay in feeding the bees from a colony suffering with an infection of EFB is that the honey they are carrying in their crop will be used up drawing the foundation. The delay ensures that there will be no remaining bacteria in the crop to infect young larvae as it will take at least 3 days till the first eggs hatch and require feeding. Remember EFB is a Notifiable Disease.

Spreading brood

This technique should not be attempted until you become more experienced. This is a very useful and powerful technique if used correctly. It can be used to build average colonies into larger ones, but has also a diagnostic value as a lot can be learned about the colony from its reaction to the spreading. One would not spread the brood of a colony badly affected by Nosema because the scarcity of bees might cause chilling but in any case a nosemic colony cannot be improved until the source of infection on the combs has been removed. Average colonies and ones with plenty of bees can however be treated with perfect assurance that it can only do good providing the one cardinal rule is not broken. This rule is that a broodless comb should never be introduced into the brood area. That is if there are five frames of brood it is bad practice to put an empty comb within the five, thus making the brood extend over six frames.

In the build-up period of the season the technique makes use of two facts. First that sealed brood produces about as much heat per individual as the individual brooding bees on the surface of the comb, and second that bees will not tolerate a brood nest which is not elliptical. Thus if a comb with a small area of brood is placed between two combs with larger areas of brood the bees will immediately clean up the extra cells, and the queen will lay in them, thus bring all the brood areas to the same size. During manipulations therefore a large central comb of sealed, preferably emerging, brood is selected and placed on the outside of the brood nest, against the smaller outside frame of brood. Remember the brood nest is not synonymous with the brood chamber, it is the area which contains brood. This will induce extra laying by the queen and thus increase the rate of build-up of the colony. If a comb with emerging brood is used, the brood will have emerged by the time of the next visit and therefore the vigour with which the queen takes up the extra heated comb area and relays the spread comb will give a good idea of her quality. Later in the year when the colony is full sized spreading the brood can be used to speed up manipulations and quickly assess the queen and the probability of swarming, see 'Examination During Swarming Season' page 67.

Storage of comb

Once supers have been extracted they have to be stored for the winter in such a way that the comb is not destroyed by wax moth. Some beekeepers store the super with wet comb directly out of the extractor. The honey appears to protect the comb from wax moth damage, but it does deliquesce and run down onto the floor making a mess. A polythene sheet or tray and some newspaper under the bottom box will soak up most of the mess. Other beekeepers place the supers with wet comb, in piles of up to six, back on the colonies above a crown board with the feed hole open, allowing the bees to clean them up and take any honey down into the brood chamber. Dry supers can be stored outdoors above a screen and with a second screen below the roof. Wax moth prefer the dark and are less likely to attack a stack that is open to light and in the cooler outdoors. The use of chemical moth repellent paradichlorobenzine (PDB) is no longer legal as it will contaminate the honey. Do not store the supers near insecticidal strips or where timber has been treated for wood beetle as the wax will absorb the insecticide and kill the bees when next in use. Small quantities of comb can be placed in a domestic freezer overnight. Acetic acid fumes will kill adult moths and their eggs and is useful for use with stored brood comb. A product called 'B401' previously called 'Certan' can be sprayed onto stored brood comb. The spray contains a bacterium harmless to bees and humans, but which will kill wax moth larvae.

Syrup

Sugar syrup made to feed honey bees should only be made with clean granulated white sugar (sucrose), particularly if the syrup is to be used during the winter. Dirty, raw or brown sugar is detrimental to the bees and in some cases definitely toxic, and should not be used to make bee feed. Syrup should be made in the proportions of 2 lbs sugar to 1 pint water (1 kg sugar to ½ litre water) as this has been shown to be the most effective strength. The easiest way to achieve these quantities is to take a handy container, accurately half fill it with water and then add sugar to fill and stir. Hot water will dissolve the sugar more readily than cold. The syrup should not be boiled, nor should anything be added to it. Syrup of this strength will not ferment.

Ten day inspection system

Regular inspections should be routine for all beekeepers during the active season to do the necessary work:

- ensure the colony has a laying queen;
- giving extra supers;
- assessing the health status;
- ensuring the colony has enough stores;
- checking for swarm preparations;
- keeping records on the state of the colony.

The suggested time interval between inspections is to detect colonies preparing to swarm, and is the longest time period it is safe to leave a colony, knowing that no swarm can get away in the interval. Bees should not normally swarm before a queen cell is sealed. Routine inspections can reveal occupied queen cells. Destruction of these queen cells may result in emergency queen cells being built over existing 1 to 3 day old larvae. This makes the safe interval between inspections 3 to 5 days as a larva is sealed in its cell after 6 days.

If the queens wings are clipped a swarm might issue, but cannot get away as it will not leave without the old queen. A swarm could still go with the first virgin to emerge, though probably not until she is a day old and able to fly. This extends the safe interval to the suggested 10 days. Destroying queen cells should not be regarded as a swarm control measure in itself, but does give the beekeeper time to prepare for a more suitable form of swarm control, provided the queen is clipped and no queen cells were missed during the inspection.

Uniting colonies

Two colonies can be combined together into one by the technique of uniting. This technique is used to reduce the number of colonies being kept, to get rid of poor colonies and for queen introduction. To unite two colonies the beekeeper decides which of the two queens he wishes to keep. The other colony is opened, the queen found and removed, the crown board and roof replaced and the colony left to quieten down. In the evening, when flying has ceased, the colony with the queen is gently opened. A single sheet of newspaper is placed over the top of the brood chamber and a queen excluder is placed on this to keep it down, and prevent it blowing away. A few small holes are made in the paper, in a place between the frame top bars, with a nail or the corner of the hive tool. The queenless colony's brood chamber minus the floor is gently lifted onto the excluder above the paper, and the colonies are left undisturbed for at least six days. Usually by the next morning the ground around the hive is covered with bits of paper as the bees have cut their way through to each other. They all think they are at home so when challenges are made, because of a difference in smell, they submit and no fighting occurs. Normally uniting in this way is done in the early or late season when the colonies are in a brood chamber only, or at the most no more than a brood chamber and one super each. There is usually no need or desire to unite large colonies with several full supers.

If the colonies to be united are some way apart in the apiary then either the uniting can be carried out right away and any bees that fly home will enter the nearest colony, or they can be worked across the apiary about a yard (metre) a day until they are close together so that none of the bees are lost.

Sometimes it is nuclei which are to be united and in this case the paper method is not as easily done, so these are usually united by dusting them with plain flour. One nucleus will be queenless. The queen in the other nucleus can be placed in a butler cage to protect her. The flour is usually dispensed in an ordinary flour dredger as used by cooks. The frames are taken one at a time and dusted with flour on both sides then placed into the brood chamber. By the time the two little lots of bees have cleaned up the mess they are quite friendly.

11. Honey Bee Forage Plants

Almost all plants which produce coloured flowers can, and will, be worked by the honey bees. Some will be far more productive than others, some only produce pollen and some only nectar, but all are useful. This chapter will therefore only deal with plants of considerable importance as sources of honey, plants that are good producers but of local distribution or ones of special interest to the honey bee. The flowering periods given are for the south of England and must be adjusted for other areas.

Pollination

Pollination is the process by which pollen is transferred from the anthers, the male part of the flower, where it is produced, to the stigma, the female receptive area, of the same or another flower of the same species. When the pollen is transferred from the anther to the stigma of the same flower, or another flower on the same plant, the process is called self pollination. Cross pollination occurs when the pollen is transferred between the flowers of two different plants of the same species or two different cultivars. Once the pollen has arrived on the stigma it germinates, sends a tube down to the egg cell and fertilises it thus initiating seed development. Many plants require the intervention of bees to move the pollen to its correct destination. The honey bee is an excellent pollinator as individuals visit one species of flower at a time and thus the right pollen is being carried on their visits. The value of the pollination done by the honey bee is several hundred times as valuable as its production of honey. Growers hire colonies of bees for pollination of such plants as fruit, clover, beans, rape and many decorative flowers and vegetables grown for seed production. The honey bee provides a large proportion of the pollination required by wild flowers, a proportion which increases as the populations of wild insects are reduced by the use of pesticides, herbicide and intensive cultural methods in modern agriculture. Much valuable pollination is done by the colonies of amateur beekeepers which are widely spread in towns and the countryside.

Balsam. *Impatiens glandulifera*

Balsam grows mainly around rivers and ponds. Bees work some individual plants heavily, others are ignored. It grows in quite dense patches in some areas and bees will get quite a lot of nectar from it; they also collect lots of pollen which is white and which dusts the bees. Balsam flowers in July to end of September.

Bell Heather. *Erica* species

The wild species of this genus are found on the moors and acid soils, while many gardens contain a winter flowering heather bed. The honey bee works all of them but only occasionally gets a large crop from them. The honey is not classed as 'heather honey', as this comes only from ling *Calluna vulgaris*. The resulting honey is a normal fluid honey, with a very strong almost bitter in flavour and 'port wine' in colour. Pollen loads are white to pale grey, flowering July to early September.

Ling Heather Moors

Berberis species

The genus contains many very well-known and valued garden shrubs, and is also used as a hedge plant. In some areas *Berberis vulgaris* will be found growing wild in farm hedges, as will the closely relate *Mahonia aquifolium*. The latter flowers in January to early May and is only worked during warm spells. The flowers of Berberis, which appear in May to June, are attractive to the bees which collect both nectar and pollen.

Birdsfoot-trefoil. *Lotus corniculatus*

Nearly all the members of this family are attractive to bees and need pollination by them. This particularly beautiful bright yellow species grows wild and is grown as an agricultural crop in some countries. It flowers from June to September and provides both nectar and pollen, the latter being pale brownish yellow.

Blackberry. *Rubus fruticosus agg*

Members of this species vary very considerably in size and shape and can be found everywhere there is any rough unfilled ground. On poor soils they are often the main nectar producing plant. The plants flower from June to September and are vigorously worked by the bees providing both nectar and a greyish yellow coloured pollen. The resulting honey is of good flavour, light amber in colour and tends at times to granulate rather coarsely. This can of course be corrected, by seeding if the beekeeper wishes.

Borage. *Borago officinalis*

This flower is grown as a garden plant and also as a field crop. This is a valuable crop to beekeepers, giving both a water white honey and providing light blue grey pollen. Borage flowers from June to August and is heavily worked by honey bees, bumblebees and other insects.

Broom. *Cytisus scoparius*

This plant (right) is generally distributed growing wild on acid soils and is worked by bees for its orange brown pollen; it probably does not produce much nectar, however closely related garden flowers such as *Cytisus praecox* are heavily worked for both nectar and pollen. The plants flower in April to June.

Buckwheat. *Fagopyrum esculentum*

This agricultural crop produces a reasonable amount of very strong flavoured, fairly dark honey. The small pink or white flowers appear in July and August, and are worked by the bees for the pale yellow pollen as well as nectar.

Citrus species

The many species which are grown for their fruit are major honey producers in many parts of the world. Orange honey is probably the best known and has a very delightful flavour of orange making it unique.

Clover, Red. *Trifolium pratense*

The flowers of this plant are too deep to allow the honey bee to reach the nectar at the bottom of the tube. However, where the first growth is cut off to make silage the second growth produces shorter flowers. If this second cut occurs when there is about three days over 70°F/21°C, then it produces the greatest nectar flow most beekeepers will ever contact. The honey is sweet and pure white, granulating with a rather mealy texture; the pollen is dark brown. Plants grown of the second cut flower in August.

Clover, Wild White. *Trifolium repens*

This is an almost perfect honey bees forage plant. It is the right size, it provides large quantities of both nectar and pollen and is tough enough to stand hard wear in permanent pastures. Its one fault is that it needs temperatures of about 70°F/21°C, before it begins to produce nectar in quantity. Years ago it was Britain's major honey plant but is now drastically reduced in quantity due to the use of nitrogen fertiliser. It flowers in June/July. The honey from wild white clover is pale creamy yellow and crystallises extremely hard. The pollen is brown to brownish orange.

Cotoneaster species

Cotoneasters are well-known ornamental flowering shrubs and small trees, they are all extremely attractive to bees. Flowers in May to June, providing both nectar and pollen.

Dandelion. *Taraxacum officinale*

This wild flower is well-liked and worked by the bees. In some areas there are large fields of this weed flowering in May, but there are individual plants flowering at almost any time in summer. The pollen is bright orange and brings a lot of yellow oil into the hive which stains the honey cappings yellow. The honey is pale yellow.

Eucalyptus species

Trees of this genus are major honey plants of the warmer countries, mainly Australia, but occur in many areas as large ornamentals, and if the weather is hot can produce quite a lot of nectar.

Field Bean. *Vicia faba*

Where this flower is grown as a field crop it provides the beekeeper with a crop of very delicious pale amber honey, and the bees with considerable amounts of pollen. The colour of the latter can vary from slate grey, fawn to brown or even flesh coloured. Autumn sown beans flower in about May and spring sown in June. The flower is unusual in that the honey bee works them in three ways. Some bees tackle the flower normally going in the mouth of the tube for pollen, others work for nectar through holes bored in the base of the flowers by bumble bees, while other nectar gatherers work the black spot on the stipules, at the base of the leaf petiole, which are extra floral nectaries.

Horse Chestnut. *Aesculus species*

The white species, *A. hippocastanum* flowers in May and is a common park and hedge tree and is well worked by the bees for its honey and dark red pollen. The red species, *A. carnea* is somewhat suspect as at times it narcotises bees, particularly bumble bees (Bombus species), and these are finished off by birds. Usually it does not affect the honey bee.

Ivy. *Hedera helix*

This is in most cases the last plant of the year to provide ample nectar and pollen for the bees, and which they work with great zest. It flowers in October and the many pollen loads going into the hive will be yellow-orange.

Lime. *Tilia* species

This genus contains a number of trees which, on the right soils, are extremely productive and others which, in some areas and some years, produce nectar which is poisonous to the honey bee. *T. vulgaris* and *T. platyphyllos* are safe and the honey has a delicious minty flavour and is greenish in colour. *T. petiolaris* and *T. orbicularis* have been known to be poisonous. The bee collects little pollen from lime which flowers at the end of June and July.

Ling. *Calluna vulgaris*

This is the flower from which heather honey is produced. This unique honey is dark amber in colour and forms a jelly. Its flavour and aroma is unique, the flavour is somewhat bitter. Flowers mid August to September, it produces vast amounts of pale grey pollen which is eagerly collected by the bees.

Lucerne or Alfalfa. *Medicago sativa*

This is a major honey producer in many parts of the world. It will provide large quantities of white honey when the plant is allowed to flower in warm weather.

Poppy. *Papaver rhoeas*

This flower has no nectaries so does not provide nectar but it provides masses of very dark blue, almost black, pollen which is extremely attractive to the bee and which they will find if there are any plants around.

Rape, Oilseed. *Brassica* species

Like all cruciferous plants this one is very attractive to the bees and will provide large crops of honey, given good weather and large colonies. It has the two disadvantages that it flowers early in the year from a winter sowing, from late April to early June, though spring sown rape flowers in June. Its honey granulates so readily that this can happen in the supers unless care is taken to extract the honey as soon as possible. It also provides large quantities of bright yellow pollen each year.

False Acacia (black locust in USA) Robinia. *Robinia pseudoacacia*

This tree is a major source of honey in parts of Europe. The honey is opposite to that of oilseed rape in that it is high in fructose and therefore only granulates with difficulty or not at all. The honey is very light in colour and has a delicate flavour.

Sweet Chestnut. *Castanea sativa*

This tree is a good honey producer in a warm summer, the honey is dark amber to dark brown and will crystallise with a coarse grain. It also provides the bees with masses of pale greenish yellow pollen.

Sycamore. *Acer pseudoplatanus*

This is a common tree in many areas and it produces large quantities of nectar when the weather is warm. The resulting honey is light amber with a greenish tinge and of good flavour. Provides plenty of greenish pollen as well.

Willow. *Salix* species

One of the earliest flowering trees and a boon to beekeepers, as in many cases it produces the first big pollen flow of the year. Male and female flowers are on different plants. Both sexes provide nectar but the pollen only comes from the male catkin. The pollen is yellow and is collected in great quantity.

Ling Heather

Red Clover

Ted Hooper was born in 1918. Having learned to work timber in the family business, in the Second World War Ted served in the Eighth Army in the Middle East, Italy and the D Day Landings. After the war he went to work for a commercial beefarmer in Hampshire, running around 500 colonies. With his experience in commercial beekeeping under his belt Ted gained his BBKA exams and later the National Diploma of Beekeeping and in 1962 became Essex County Beekeeping instructor at Writtle College. Ted was subsequently active in the BBKA husbandry commitee for many years. He was elected to the BBKA Executive in 1978 and was chairman in 1989, then president in 1991. He also served on the BBKA exam board for eighteen years, chairing it from 1972 to 1988. His work in the field of beekeeping was recognised with the award of an MBE. Besides *The Beginner's Bee Book*, Ted was also the author of the bestselling *Guide to Bees and Honey* which has sold over 100,000 copies worldwide and *The Illustrated Encyclopaedia of Beekeeping* with Roger Morse and *The Beekeeper's Garden* (later retitled as *The Bee Friendly Garden*) with Mike Taylor. Ted died in March 2010, aged 91.

Clive de Bruyn was born in 1943 and with an honours degree in metallurgy spent his early working years in the Sheffield steel industry. He started keeping bees in the 1960s and gained his National Diploma in Beekeeping in 1976. Shortly afterwards he joined the National Bee Unit where he stayed for four years before joining Honey Farmers Ltd, a large national commercial beekeeping company, as a trouble shooter. From 1978 to 1982 he was employed by MAFF as a Beekeeping Adviser at the National Beekeeping Unit in charge of bee diseases. From 1983 to 1997 Clive was County Beekeeping Instructor for Essex, having been appointed on Ted Hooper's retirement. Clive has been a self-employed consultant and professional beekeeper from 1997 to the present and continues to manage 200 colonies. He has travelled extensively abroad to see how beekeeping is practised in other countries and has written numerous papers and articles as well as two books *Practical Beekeeping* and *The New Varroa Handbook* with Bernard Mobus.

Margaret Thomas and her husband started keeping bees in 1973 and studied beekeeping under Ted Hooper at Writtle College, passing her BBKA examinations and qualifying as a Master Beekeeper. She gained the National Diploma in Beekeeping in 1982. Suburban beekeeping expanded into numerous out-apiaries on Essex farms and active membership of the Bee Farmers Association for whom she was Magazine Secretary, Spray Delegate and Pollination Secretary. She has served on the BBKA Examinations Board as Moderator and tutors for the BBKA correspondence course. Having managed 60 colonies in Essex, Margaret and her husband moved to Aberfeldy in 2008. They now manage five colonies and are relearning beekeeping in the harsher, Scottish climate. Margaret continues to be involved in training programmes in Scotland through the Scottish Beekeepers Association.